Lecture Notes in Business Information Processing 10

Jan L.G. Dietz Antonia Albani
Joseph Barjis (Eds.)

Advances in Enterprise Engineering I

4th International Workshop CIAO! and
4th International Workshop EOMAS, held at CAiSE 2008
Montpellier, France, June 16-17, 2008
Proceedings

Volume Editors

Jan L.G. Dietz
Antonia Albani
Delft University of Technology
Mekelweg 4, 2628 CD Delft, The Netherlands
E-mail: {j.l.g.dietz,a.albani}@tudelft.nl

Joseph Barjis
University of Wisconsin
2100 Main St., Stevens Point, WI 54481, USA
E-mail: jbarjis@uwsp.edu

Library of Congress Control Number: 2008928014

ACM Computing Classification (1998): J.1, D.2, H.4

ISSN 1865-1348
ISBN-10 3-540-68643-6 Springer Berlin Heidelberg New York
ISBN-13 978-3-540-68643-9 Springer Berlin Heidelberg New York

Springer is a part of Springer Science+Business Media

springer.com

Printed in Germany

Typesetting: Camera-ready by author, data conversion by Scientific Publishing Services, Chennai, India
Printed on acid-free paper SPIN: 12275977 06/3180 5 4 3 2 1 0

Preface

The expectation for the future of the 21st century enterprise is complexity and agility. In this digital age, business processes are scattered not only throughout the labyrinth of their own enterprises, but also across different enterprises, and even beyond the national boundaries. An evidence of this is the growing phenomenon of business process outsourcing. Increasing competition, higher customer demands, and emerging technologies require swift adaptation to the changes.

To understand, design, and engineer a modern enterprise (or an enterprise network) and its interwoven business processes, an engineering and systematic approach based on sound and rigorous theories and methodologies is necessary. Along with that, a paradigm shift seems to be needed for addressing these issues adequately. An appealing candidate is to look at an enterprise and its business processes as a social system. In its social setting, an enterprise and its business processes represent actors with certain authorities and assigned roles, who assume certain responsibilities in order to provide a service to its environment.

The need for this paradigm shift along with the complexity and agility of modern enterprises, gives inspiration for the emerging discipline of *Enterprise Engineering*. For the study of this socio-technical phenomenon, the prominent tools of *Modeling* and *Simulation* play a significant role. Both (conceptual) modeling and simulation are widely used for understanding, analyzing, and engineering an enterprise (its organization and business processes).

In addressing the current challenges and laying down some principles for enterprise engineering, this book includes a collection of papers presented and discussed at the joint meeting of CIAO! 2008 and EOMAS 2008, organized in conjunction with the 20th CAiSE conference. The scopes of these two workshops are to a large extent complementary, with CIAO! being more focused on the theory and application of enterprise engineering and EOMAS on the methods and tools for modeling and simulation.

June 2008

Jan L.G. Dietz
Antonia Albani
Joseph Barjis

Enterprise Engineering – A Manifesto

Traditional organizational sciences fall short in assisting enterprises to adapt their strategies and to implement them effectively and flexibly. Enterprise Engineering is a design-oriented approach for coping with these problems, based on the merging of organizational sciences and information system sciences. It aims at bringing rigor to organizational design.

Motivation

The prosperity of modern society is largely determined by the societal performance of enterprises, of all kinds, including commercial, non-profit, and governmental companies and institutions, as well as all kinds of alliances between them, such as virtual enterprises and supply chains. By the societal performance of an enterprise is understood its performance in all aspects that are considered important, notably economic, social, and ethical. Ultimately, it depends on two key factors. One is the strategy chosen by the enterprise, as well as the continuous adaptation of this strategy to upcoming threats and challenges. The other one is the implementation of this strategy in a comprehensive, coherent, consistent, as well as efficient and adaptable way.

Unfortunately, the vast majority of strategic initiatives fail, meaning that enterprises are unable to gain success from their strategy. The high failure rates are reported from various domains: total quality management, business process reengineering, six sigma, lean production, e-business, customer relationship management, as well as from mergers and acquisitions. Whereas often, presumably for convenience sake, unforeseen or uncontrollable events are presented as the causes of failure, research has shown that strategic failure is mostly the avoidable result of inadequate strategy implementation. Rarely is it the inevitable consequence of a poor strategy.

A plethora of literature indicates that the key reason for strategic failures is the lack of coherence and consistency, collectively also called congruence, among the various components of an enterprise. At the same time, the need to operate as an integrated whole is becoming increasingly important. Globalization, the removal of trade barriers, deregulation, etc., have led to networks of cooperating enterprises on a large scale, enabled by the virtually unlimited possibilities of modern information and communication technology. Future enterprises will therefore have to operate in an ever more dynamic and global environment. They need to be more agile, more adaptive, and more transparent. In addition, they will be held more publicly accountable for every effect they produce.

Said problems are often addressed with black-box thinking based knowledge, i.e., knowledge concerning the function and the behavior of enterprises. Such knowledge is sufficient, and perfectly adequate, for managing an enterprise within the current range of control. However, it is fundamentally inadequate for changing an enterprise, which is necessary to meet performance goals that are outside the current range of control. In order to bring about those changes in a systematic and controlled way, white-box based knowledge is needed, i.e., knowledge concerning the construction and the operation of enterprises. Developing and applying such knowledge requires no less than a paradigm shift in our thinking about enterprises, since the traditional organizational sciences are dominantly oriented towards organizational behavior, based on black-box thinking.

The needed new paradigm is that enterprises are purposefully designed systems. The needed new skill is to (re)design, (re)engineer, and (re)implement an enterprise in a comprehensive, coherent and consistent way (such that it operates as an integrated whole), and to be able to do this whenever it is needed.

The Paradigm Shift

The current situation in the organizational sciences resembles very much the one that existed in the information systems sciences around 1970. At that time, a revolution took place in the way people conceived information technology and its applications. Since then, people have been aware of the distinction between the *form* and the *content* of information. This revolution marks the transition from the era of Data Systems Engineering to the era of Information Systems Engineering. The comparison we draw with the computing sciences is not an arbitrary one. On the one hand, the key enabling technology for shaping future enterprises is the modern information and communication technology (ICT). On the other hand, there is a growing insight in the computing sciences that the central notion for understanding profoundly the relationship between organization and ICT is the entering into and complying with commitments between social individuals. These commitments are raised in communication, through the so-called *intention* of communicative acts. Examples of intentions are requesting, promising, stating, and accepting. Therefore, as the content of communication was put on top of its form in the 1970's, the intention of communication is now put on top of its content. It explains and clarifies the organizational notions of collaboration and cooperation, as well as authority and responsibility. This current revolution in the information systems sciences marks the transition from the era of Information Systems Engineering to the era of Enterprise Engineering, while at the same time merging with relevant parts of the Organizational Sciences, as illustrated in Fig. 1.

Mission

The mission of the discipline of Enterprise Engineering is to combine (relevant parts from) the traditional organizational sciences and the information systems sciences, and to develop emerging theories and associated methodologies for

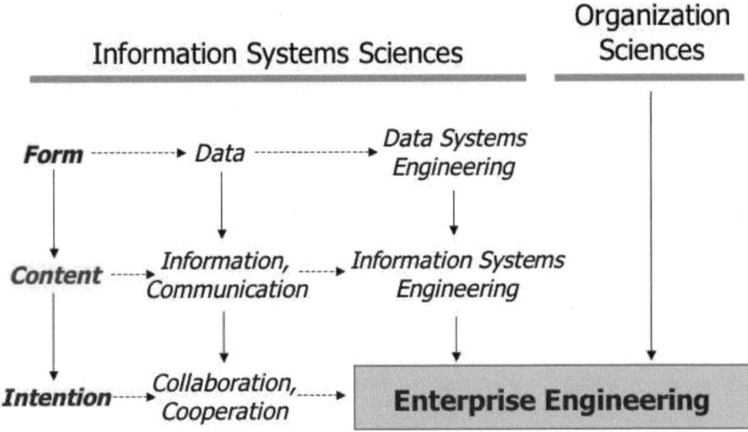

Fig. 1. Enterprise Engineering

the analysis, design, engineering, and implementation of future enterprises. Two fundamental notions have already emerged and seem to be indispensable for accomplishing this mission: Enterprise Ontology and Enterprise Architecture.

Enterprise Ontology is conceptually defined as the understanding of an enterprise's construction and operation in a fully implementation-independent way. Practically, it is the highest-level constructional model of an enterprise, the implementation model being the lowest one. Compared to its implementation model, the ontological model offers a reduction of complexity of well over 90%. It is only by applying this notion of Enterprise Ontology that substantial strategic changes of enterprises can be made intellectually *manageable*.

Enterprise Architecture is conceptually defined as the normative restriction of design freedom. Practically, it is a coherent and consistent set of principles that guide the design, engineering, and implementation of an enterprise. Any strategic initiative of an enterprise can only be made operational through transforming it into principles that guide the design, engineering, and implementation of the new enterprise. Only by applying this notion of Enterprise Architecture can *consistency* be achieved between the high-level policies (mission, strategies) and the operational business rules of an enterprise.

June 2008

Jan L.G. Dietz
Jan A.P. Hoogervorst

Organization

The CIAO! and EOMAS workshops are organized annually as two international forums for researchers and practitioners in the general field of Enterprise Engineering. Organization of these two workshops and peer review of the contributions made to these workshops are accomplished by an outstanding international team of experts in the fields of Enterprise Engineering, Modeling and Simulation.

Workshop Chairs

CIAO! 2008

Jan L.G. Dietz	Delft University of Technology (The Netherlands)
Antonia Albani	Delft University of Technology (The Netherlands)

EOMAS 2008

Joseph Barjis	University of Wisconsin - Stevens Point (USA)

Program Committee

CIAO! 2008

Wil van der Aalst
Bernhard Bauer
Johann Eder
Joaquim Filipe
Rony G. Flatscher
Kees van Hee
Birgit Hofreiter
Jan Hoogervorst
Emmanuel dela Hostria
Christian Huemer
Zahir Irani
Peter Loos
Graham Mcleod
Arturo Molina
Aldo de Moor
Hans Mulder
Moira Norrie
Andreas L. Opdahl
Maria Orlowska
Martin Op 't Land
Erik Proper
Gil Regev
Dewald Roode
Pnina Soffer
José Tribolet
Johannes Maria Zaha

EOMAS 2008

Anteneh Ayanso
Manuel I. Capel-Tuñón
Rodney Clarke
Ashish Gupta

Oleg Gusikhin
Selma Limam Mansar
Mikael Lind
Prabhat Mahanti
Yuri Merkuryev
Vojtěch Merunka
Alta van der Merwe
Murali Mohan Narasipuram
Oleg V. Pavlov
Viara Popova
Srini Ramaswamy
Han Reichgelt
Peter Rittgen
Natalia Sidorova
José Tribolet
Alexander Verbraeck
Gerald Wagner

Sponsoring Organizations

- CAiSE 2008 (International **C**onference on **A**dvanced **I**nformation **S**ystems Engineering)
- SIGMAS (**S**pecial **I**nterest **G**roup on **M**odeling **A**nd **S**imulation of the Association for Information Systems)

Table of Contents

Process Modeling

Collaboration and Interoperability

Enterprise Architecture

Model Transformation and Simulation

On the Nature of Business Rules

Jan L.G. Dietz

Delft University of Technology, the Netherlands
j.l.g.dietz@tudelft.nl

Abstract. Business rules are in the center of attention, both in the 'business world' and in the 'ICT applications world'. Recently, the OMG has completed a major study in defining the notion of business rule and its associated notions. On closer look, however, the definitions provided appear to be not as rigid and precise as one would hope and as deemed necessary. Based on the consistent and coherent theoretical framework of Enterprise Ontology, several clarifications of the core notions regarding business rules are presented. They are illustrated by means of a small example case.

Keywords: Business Rule, Enterprise Ontology, DEMO, Modal Logic.

1 Introduction

1.1 A Survey of Current Business Rule Notions

Business rules constitute a subject of topical interest. They are presented and promoted as a means to achieve several highly valued properties of information systems (ICT applications), like flexibility, maintainability, transparency, and cost savings. Recently, the Object Management Group (OMG) adopted the SBVR standard (Semantics of Business Vocabulary and Business Rules) for specifying business objects, facts, and rules [13]. However, even in this impressive piece of work, the core notions appear not to be defined as crisply as one would whish.

One of the most well known documents regarding business rules is the authoritative work of Ronald Ross [14]. According to Ross, business rules build on terms and facts. A term is a basic noun or noun phrase in natural language. Examples of terms (taken from [14]) are:

Customer	(Basic? Atomic?)
Order	(Atomic?)
Quantity back-ordered	(Basic?)
Employee name	(Knowable?)

In order to keep the set of terms manageable, Ross proposes three fundamental tests that terms have to pass in order to be included. First, they should represent the most *basic* things of an enterprise, i.e., they cannot be derived or computed. Second, they should be *atomic*, i.e., they should represent things that are indivisible. Third, they should be *knowable*, i.e., they should represent things that exist, rather than things hat happen. Unfortunately, no hard criteria are provided for determining whether a term

J.L.G. Dietz et al. (Eds.): CIAO! 2008 and EOMAS 2008, LNBIP 10, pp. 1–15, 2008.

succeeds or fails to pass the tests. Anticipating on the results of applying the Enterprise Ontology Theory, as presented in Section 3, we have put already some question marks next to the example list of terms above; they will be addressed later. Facts, according to Ross, are expressed by sentences that follow the subject-verb-object structure, where subjects and objects are referred to by terms. Examples of facts (taken from [14]) are:

Customer *places* order	(verbal predicate))
Order *is included in* shipment	(nominal predicate)
Employee *has a* gender	(nominal predicate)
Manager *is a category of* employee	(instance of meta fact type)

Again anticipating on the results of applying the Enterprise Ontology Theory in Section 3, the first observation to be made is that apparently fact types are represented: the sentences that can be instantiated. The fourth sentence is an exception. It cannot be instantiated since it is itself an instance, be it of a meta fact type. Second, no distinction is made between nominal and verbal predicates, i.e., between facts and (apparent) acts.

In accordance with many other authors (e.g., [10], [12]), Ross requires business rules to be declarative, instead of procedural, and to be expressed in well-formed (ideally: logical) formulas. Three fundamental categories are distinguished: rejectors, producers, and projectors. A *rejector* is a rule that constraints the behavior of a business. Rejectors can be violated. A *producer* is a rule for computing or logical derivation. A *projector* is a rule through which an action is evoked. Examples of each of these categories are:

A customer cannot rent more than one car at the same time	(rejector)
The amount to be paid is the list price plus VAT	(producer)
Reorder stock if the quantity on hand drops below some level	(projector)

Distinguishing between categories of rules undoubtedly makes sense but some questions immediately pop up: Why three? Why these three? Halpin [10], for example, proposes a subdivision of the rejectors in static and dynamic constraints, a distinction that is founded in database research. Next, the formulation of the projector example can hardly be called declarative. So, this contradicts the point of departure.

1.2 Research Questions and Research Approach

The short survey above should suffice to sketch the problem area we want to address and to formulate research questions that have both societal and scientific relevance, being motivated by the conviction that conceptual frameworks, like the SBVR, should be made much more rigid. The research questions to be addressed are:

1. Business rules appear to support and guide the operations of a business. But what is exactly their scope? In particular, how are they distinguished from design principles, as incorporated in the notion of architecture?
2. How can the notion of business rule be made crisper? Related to that, how important is the way of formulation (declarative-shape or imperative-shape)?

3. How is the notion of business rule related to the notions of business object, business fact, and business event?
4. What useful distinctions can be made in order to keep the total set of business rules manageable? Related to this: what makes a rule a business rule?

We will seek answers to these questions on the basis of a scientifically sound foundation, namely Enterprise Ontology, in particular its underlying Ψ-theory [7]. The Ψ-theory offers a coherent and consistent understanding of the operation of an enterprise. Such a theory is the basis one needs to clearly and precisely define core notions like (business) rules, (business) objects, (business) facts, and (business) events. Any other basis will at best reduce the current confusion to some extent, but not sufficiently. The ambition of the research project on which this paper reports, is to remove the confusion definitively.

In Section 2, the theoretical basis of our research approach is summarized. Space limitations force us to keep it rather concise which means that a reader who is totally unfamiliar with the notion of Enterprise Ontology may need to read some references. On the basis of the presented theory, we will clarify the notion of business rule, as well as related notions, in Section 3. The analysis is illustrated by a small example case. Section 4 contains the conclusions that can be drawn.

2 An Introduction to Enterprise Ontology[1]

2.1 Theoretical Foundations

There exist two different system notions, each with its own value, its own purpose, and its own type of model: the function-oriented or teleological and the construction-oriented or ontological system notion [2]. The *teleological system* notion is about the function and the (external) behavior of a system. The corresponding type of model is the *black-box model*. Ideally, such a model is a (mathematical) relation between a set of input variables and a set of output variables, called the transfer function. The teleological system notion is adequate for the purpose of using or controlling a system. It is therefore the dominant system concept in e.g. the social sciences, including the organizational sciences. For the purpose of building and changing a system, one needs to adopt the ontological system notion. It is therefore the dominant system notion in all engineering sciences.

The *ontological system* notion is about the construction and operation of a system. The corresponding type of model is the *white-box model*, which is a direct conceptualization of the ontological system definition presented below. The relationship with function and behavior is that the behavior is brought forward, and consequently explained, by the construction and the operation of a system. These definitions are in accordance with the work of Gero et al. if one substitutes their use of "structure" by "construction and operation" [9]. The ontological definition of a system, based on the one that is provided in [2], is as follows. Something is a system if and only if it has the next properties:

[1] The contents of this section is based on the Ψ-theory [7]. The Greek letter Ψ is pronounced as PSI, which stands for Performance in Social Interaction. It constitutes the basic paradigm of the theory and conveys the underlying philosophical stance of constructivism [15].

- *Composition*: a set of elements of some category (physical, biological, social, chemical etc.).
- *Environment*: a set of elements of the same category. The composition and the environment are disjoint.
- *Structure*: a set of influencing bonds among the elements in the composition and between these and the elements in the environment.
- *Production*: the elements in the composition produce services that are delivered to the elements in the environment.

Associated with every system is the world in which the actions of the system get their effect. The *state* of a world is a set of facts. The *state space* of a world is the set of lawful states, and the *transition space* is the set of lawful sequences of transitions. The occurrence of a transition is called an *event.*

A *fact* is something that is the case [17]. The knowledge of a fact can be expressed in a predicate over one or more objects, where an object is conceived as a bare individual [1]. We will consider only elementary facts [6, 10]. Facts can be declared, like the declaration of the concept 'car', or defined, like the definition of the concept 'van' on the basis of the concept 'car'. This notion of fact is all one needs for modeling a world. It is only a matter of convenience to conceive of *entities* next to facts. An entity type is just a unary fact type, for example the type car. Including both types and classes in a conceptual model is also a matter of convenience. An entity *class* is just the extensional counterpart of the (intensional) entity type. As an example, the class CAR = {x | car(x)}. According to the distinction between function and construction, the collective services provided by an enterprise to its environment are called the *business* of the enterprise; it represents the function perspective. Likewise, the collective activities of an enterprise in which these services are brought about and delivered, including the human actors that perform these activities, are called the *organization* of the enterprise; it represents the construction perspective. An organization is a system in the category of social systems. This means that the elements are social individuals, i.e. human beings or subjects in their ability of entering into and complying with commitments about the things that are produced in cooperation. Subjects fulfill actor roles (to be explained later). A subject in its fulfillment of an actor role is called an *actor.*

2.2 The Universal Transaction Pattern

Actors perform two kinds of acts. By performing *production acts*, the actors contribute to bringing about and delivering services to the environment of the organization. A production act (P-act for short) may be material (manufacturing, transporting, etc.) or immaterial (deciding, judging, diagnosing, etc.). By performing *coordination acts* (C-acts for short), actors enter into and comply with commitments. In doing so, they initiate and coordinate the performance of production acts. Examples of C-acts are requesting and promising a P-fact. The result of successfully performing a C-act is a *coordination fact* or C-fact (e.g., the being requested of a P-fact).

The result of successfully performing a P-act is a *production fact* or P-fact. P-facts in the case Library (see Sect. 3) are "loan L has been started" and "the late return fine for loan L has been paid". The variable L denotes an instance of loan. An *actor role* is defined as a particular, atomic 'amount' of authority, viz. the authority needed to perform precisely one kind of production act.

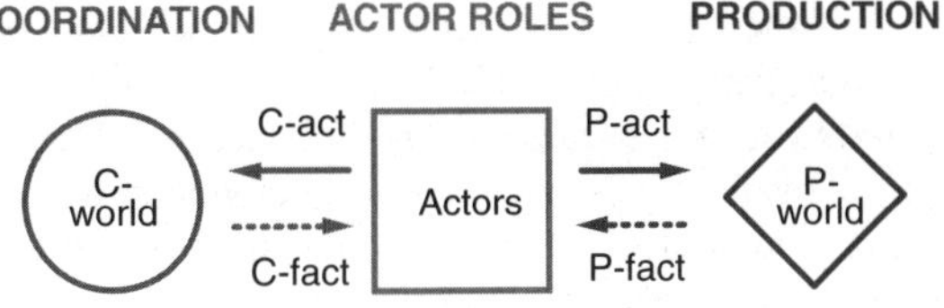

Fig. 1. The white-box model of an organization

Just as we distinguish between P-acts and C-acts, we also distinguish between two worlds in which these kinds of acts have effect: the *production world* or P-world and the *coordination world* or C-world respectively (see Fig. 1). At any moment, the C-world and the P-world are in a particular state, simply defined as a set of C-facts or P-facts respectively. When active, actors take the current state of the P-world and the C-world into account (indicated by the dotted arrows in Fig. 1). C-facts serve as agenda for actors, which they constantly try to deal with. Otherwise said, actors interact by means of creating and dealing with C-facts. The *operational principle* of organizations is that actors feel committed to deal adequately with their agenda.

P-acts and C-acts appear to occur in generic recurrent patterns, called *transactions* [4, 7]. Our notion of transaction is to a some extent similar to the notion of Conversation for Action in [16] and to the notion of Workflow Loop in [3]. A transaction goes off in three phases: the order phase (O-phase), the execution phase (E-phase), and the result phase (R-phase). It is carried through by two actors, who alternately perform acts. The actor who starts the transaction and eventually completes it, is called the initiator or *customer*. The other one, who actually performs the production act, is called the executor or *producer*. The O-phase is a conversation that starts with a request by the customer and ends (if successfully) with a promise by the producer. The R-phase is a conversation that starts with a statement by the producer and ends (if successfully) with an acceptance by the customer. In between these two conversations there is the E-phase in which the producer performs the P-act.

In Fig. 2, we present the basic form of this transaction pattern. It shows that the bringing about of an original new, thus, ontological, production result (as an example: the delivery of a bouquet of flowers) starts with the requesting of this result by someone in the role of customer from someone in the role of producer. The original new thing that is

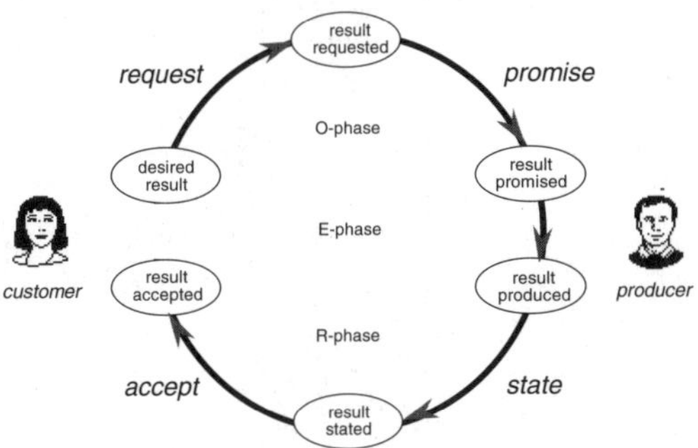

Fig. 2. The basic pattern of a transaction

created by this act, as is the case for every coordination act, is a *commitment*. Carrying through a transaction is a "game" of entering into and complying with commitments.

So, the process starts with the request for the bouquet by the customer, which brings the process to the state "result requested", the result being the ownership by the customer of the desired bouquet. The producer responds to the state "result requested" by promising to bring about the desired result, which brings the process to the state "result promised". This represents a to-do item for the producer: he has to comply with the promise by actually delivering the bouquet of flowers, i.e., executing the production act. In the act of handing over the bouquet to the customer, he states that he has complied with his promise. The process now comes to the state "result stated". The customer responds to this state by accepting the result. This act completes the transaction successfully.

The basic pattern must always be passed through for establishing a new P-fact. A few comments are in place however. First, performing a C-act does not necessarily mean that there is oral or written communication. Every (physical) act may count as a C-act. Second, C-acts may be performed *tacitly*, i.e. without any signs being produced. In particular the promise and the acceptance are often performed tacitly (according to the rule "no news is good news"). Third, next to the basic transaction pattern, as presented in Fig. 2, two dissent patterns and four cancellations patterns are identified [4, 7]. Together with the standard pattern they constitute the complete transaction pattern. It is exhibited in Fig. 3. Next to the basic transaction steps (a step is a combined C-act and C-fact) discussed before, there is the decline as the alternative of a promise, and the reject as the alternative of an accept. Both C-facts are discussion states, where the two actors have to 'sit together' and try to come to a (new) agreement. When unsuccessful, the transaction is stopped, either by the initiator or by the executor. Four cancellation patterns, on the left and the right side, complete the transaction pattern, one for every basic step.

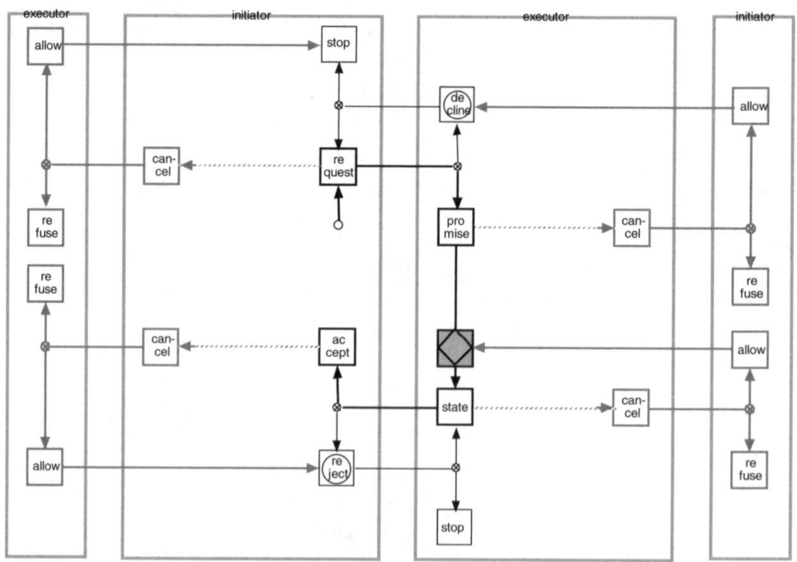

Fig. 3. The universal transaction pattern

Every *transaction process* is some path through this complete pattern, and every *business process* in every organization is a connected collection of such transaction processes. This holds also for processes across organizations, like in supply chains and networks. That is why the transaction pattern is universal and must be taken as a *socionomic law*: people always and everywhere conduct business (of whatever kind) along this pattern [7].

2.3 The Aspect Organizations

Three human abilities play a significant role in performing C-acts. They are called forma, informa and performa respectively [7]. The *forma* ability concerns being able to produce and perceive sentences. The *informa* ability concerns being able to formulate thoughts into sentences and to interpret sentences. The term 'thought' is used in the most general sense. It may be a fact, a wish, an emotion etc. The *performa* ability concerns being able to engage into commitments, either as performer or as addressee of a coordination act. This ability may be considered as the *essential* human ability for doing business (of any kind).

From the production side, the levels of ability may be understood as 'glasses' for viewing an organization (see Fig. 4). Looking through the *ontological* glasses, one observes the business actors (B-actors), who perform P-acts that result in original (i.e., non-derivable) facts. So, an ontological act is an act in which new original things are brought about. Deciding and judging are typical ontological production acts. Ontological production acts and facts are collectively called B-things. Looking through the *infological*[2] glasses, one observes intellectual actors (I-actors), who perform infological acts like deriving, computing, and reasoning. As an example, calculating the late return fine in the case Library (Sect. 3) is an infological act. Infological production acts and facts are collectively called I-things.

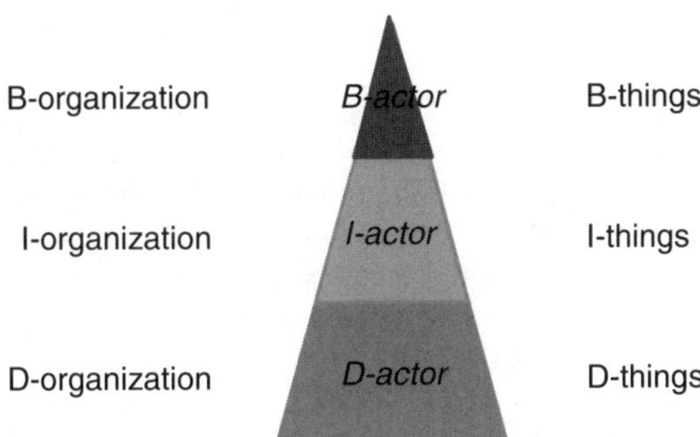

Fig. 4. Depiction of the aspect organizations

Looking through the *datalogical* glasses, one observes datalogical actors (D-actors), who execute datalogical acts like gathering, distributing, storing, and copying documents containing the facts mentioned above. So, a datalogical production act is an act in which one manipulates the form of information, commonly referred to as data, without

[2] The notions "infological" and "datalogical" are taken from Langefors [11].

being concerned about its content. For example, the act of recording a loan in the Library's database is a datalogical act. Datalogical production acts and facts are collectively called *D-things*.

The distinction levels as exhibited in 4 are an example of a *layered nesting* of systems [2]. Generally spoken, the system in some layer *supports* the system in the next higher layer. Conversely, the system in some layer *uses* the system in the next lower layer. So, the B-organization uses the I- organization and the I- organization uses the D- organization. Conversely, the D- organization supports the I- organization and the I- organization supports the B- organization.

In the Ψ-theory based DEMO methodology[3], four aspect models of the complete ontological model of an organization are distinguished, as exhibited in 5. The Construction Model (CM) specifies the construction of the organization: the actor roles in the composition and the environment as well as the transaction types in which they are involved. The Process Model (PM) specifies the state space and the transition space of the C-world. The State Model (SM) specifies the state space and the transition space of the P-world. The Action Model consists of the action rules that serve as guidelines for the actor roles in the composition of the organization.

Enterprise Ontology is one of the two pillars of the emerging field of Enterprise Engineering, Enterprise Architecture being the other one [8]. The paradigm of Enterprise Engineering is that an enterprise[4] is a designed artifact. Its implication is that any change of an enterprise, however small, means a redesign of the enterprise, mostly only a redesign of its construction, sometimes also a redesign of its function.

3 Assessing the Notion of Business Rule

3.1 Clarifications

As we have seen in Sec. 1, the core notion of business rule, common to all sources, is that it is a constraint on the behavior of an enterprise; it specifies what is allowable and what isn't. Within Enterprise Engineering, a clear distinction is made between the design phase and the operational phase of an enterprise [8]. In the design phase, one is concerned with the (re)design of both the function of the enterprise (its business) and the construction (its organization). The design process is guided by the applicable functional and constructional design principles. They are the operationalization of the notion of Enterprise Architecture, as explained in [8].

The *first clarification* we propose is to use the term "business rule" exclusively for the operational phase of an enterprise, thus to consider business rules as operational rules. Consequently, business rules are determined during the (re)design of an enterprise, and every change of a business rule, as well as the addition or deletion of a rule, implies a redesign of the enterprise. The relationship between business rules and enterprise policies is therefore indirect, namely via the design principles that are applied in the design phase. The knowledge sources we have referred to in Sec. 1 don't contain explicit statements regarding the distinction between design phase and operational

[3] Design and Engineering Methodology of Organizations, see www.demo.nl

[4] We use the term "enterprise" in a most general way. It refers to companies, to governmental agencies, to unions, to not-for-profit institutions, etc.

phase. This prevents us from elaborating the issue, conjecturing at the same time that such a strict distinction is not made.

On the basis of the holistic enterprise ontology, as discussed in Sec. 2, the *second clarification* we propose is to consider business rules as specifications of the state space and the transition space (of both the production world and the coordination world) of an enterprise's B-organization.

Although business rules can very well be expressed in natural language [14] and in diagrammatic languages [10], the most precise and concise way is to express them by formulas in modal logic [7, 13]; conversely, one could say that it is the nature of a business rule to be a formula in modal logic. A modal logic formula is a (first-order, all-quantified) logical formula preceded by a modal operator. It appears that one can distinguish between two modal operators: *necessity* (with its negation: possibility) and *obligation* (with its negation: prohibition). Let us, based on the two modal operators, distinguish between declarative-shape and imperative-shape business rules respectively. A declarative-shape business rule expresses a constraint on the state space or the transition space of a world. Examples of a state space constraint and a transition space constraint are respectively:

A customer cannot rent more than one car at the same time
The start of a car rental has to be preceded by depositing a particular amount

An imperative-shape business rule expresses how to respond to a business event, like a procedure or protocol. It is important to notice that imperative-shape business rules do not come in addition to declarative-shape business rules but that they are operational transformations of declarative-shape business rules. Applying the imperative-shape business rules of an organization guarantees that one is compliant with the declarative-shape business rules. So, strictly spoken, one can do with only imperative-shape business rules. However, providing also the declarative-shape business rules gives much more insight in the state space and the transition space of both the production and the coordination world. Next, it is a helpful intermediate stage in formulating imperative-shape business rules. Of course, one has to take care that the imperative-shape rules and the declarative-shape rules are mutually consistent.

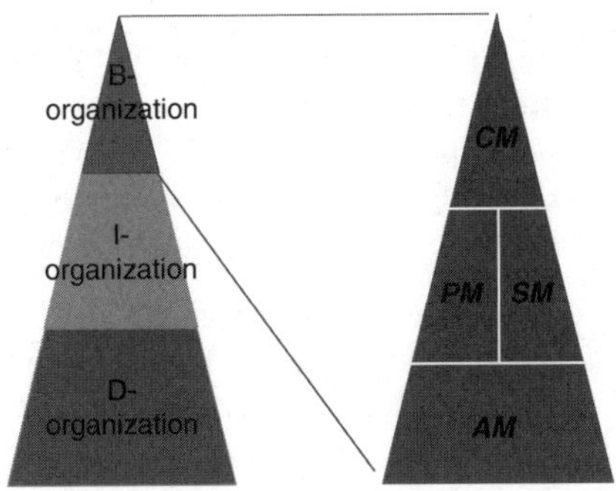

Fig. 5. The ontological aspect models

Projecting the modal operators on the four aspect models of DEMO (Fig. 5), it turns out that the declarative-shape business rules are contained in the State Model and the Process Model, and the imperative-shape business rules in the Action Model. Thus, the latter ones are action rules, prescribing how actors should respond to business events. As said before, declarative-shape business rules are constraints on the state space and the transition space of both the C-world and the P-world. Regarding the P-world, we propose to call its state elements business facts, being predications over business objects. Regarding the C-world, we already called C-facts business events. This is the *third clarification* we propose.

The last and *fourth clarification* we propose is to distinguish between the three aspect organizations. This separation of concerns is very useful in making the total set of 'business' rules manageable, next to the adoption of the universal transaction pattern that contains already a bunch of predefined rules. Both Halpin's categories of static constraints and dynamic constraints [10] and Ross' rejectors [14] presumably cover all three aspect organizations.

Next, we propose to reserve the term "business rule" exclusively for B-organization rules. This position is contrary to OMG's [13], where a business rule is defined as a rule under business jurisdiction. This doesn't seem to be a good criterion. First, it obfuscates that enterprises are also subject to rules from outside, e.g., from national legislation. Second, naming a concept by a term that does not reflect an inherent property of the concept is never a good idea, since the essence of the concept will not be captured.

3.2 Illustrations

To elaborate and illustrate our point of view, let us take an example case, namely the case Library [5, 7]. A general understanding of the operations of a library is sufficient to keep up with the discussion. We will focus on the processes concerning book loans. Among others, the next constraints apply:

A member cannot lend more than max_copies_in_loan at the same time (1)
Lent books have to be returned within the standard_loan _period (2)
A loan cannot be ended if the book copy has not been returned (3)
Loans that are ended too late will be fined with the incurred_fine amount (4)
A person may have more than one membership at the same time (5)

Let us make some preliminary observations. Rule 1, 3, and 5 are state space constraints. Rules 2 and 4 are transition space constraints. Note that all rules can be violated except rule 5. Let us next project these rules on the theoretical foundations of Enterprise Ontology. Without further explanation we state that the next transactions, including their results, are involved in these rules, taken from [5, 7]:

T04 loan start	R04 loan L has been started
T05 book return	R05 the book copy of loan L has been returned
T06 loan end	R06 loan L has been ended
T07 fine payment	R07 the late return fine for loan L has been paid

The corresponding parts of the Process Model of the Library are exhibited in Fig. 6 and Fig. 7.

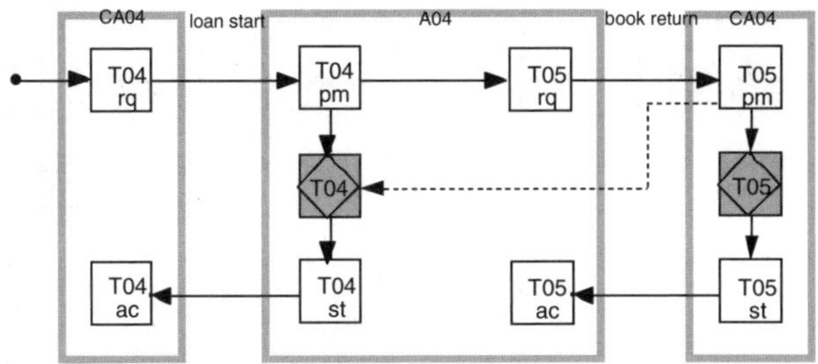

Fig. 6. Process Model of the loan start process

A box (symbol for an act) including a disk (symbol for coordination) represents a C-act (e.g., request T04) and its resulting C-fact (e.g., T04 requested), collectively called a transaction step, and indicated with e.g., T04/rq[5]. A box including a diamond (symbol for production) represents a P-act (e.g., the P-act of T04) and its resulting P-fact (e.g., the P-fact of T04), collectively called the execution step of the transaction, and indicated with e.g., T04. A solid arrow from a step S1 to a step S2 expresses the constraint that S1 precedes S2 as well as that S2 has to be performed after S1. So, for example, the promise of T04(L), where L denotes some loan, precedes the request of T05(L), and this request has to be performed once T04(L) is promised. A dotted arrow from S1 to S2 expresses only the constraint that S2 precedes S1. So, for example, the promise of T05(L) precedes the execution of T04(L), i.e., actor A04 has to wait for executing T04(L) until T05(L) is performed. The gray-lined rectangles represent the responsibility areas of the involved actor roles. For example, A04 is responsible for performing T04/pm, T05/rq, T04/ex, T04/st, and T05/ac. Note that, for the sake of simplicity, only the basic transaction pattern is shown in Fig. 6 and Fig. 7.

The corresponding, complete, part of the Action Model for actor A04 consists of the next action rules:

When T04(L) is requested, it must be declined if the total number of books in loan under the same membership as the one for L is equal to the current maximum number of copies in loan; otherwise it must be promised.
When T04(L) is promised, T05(L) must be requested.
When T05(L) is promised, T04(L) must be executed and stated.
When T05(L) is stated, it must be rejected if the book copy is damaged; otherwise it must be accepted.

Note that we have solved the having to wait until T05(L) is promised before being able to execute T04(L) by the 'trick' that T04(L) must be executed *when* T05(L) is promised. This is fully acceptable in practice, while preserving that we are dealing with an inter-transaction relationship.

---

[5] rq stands for request, pm for promise, st for state, ac for accept, and ex for executing the P-act.

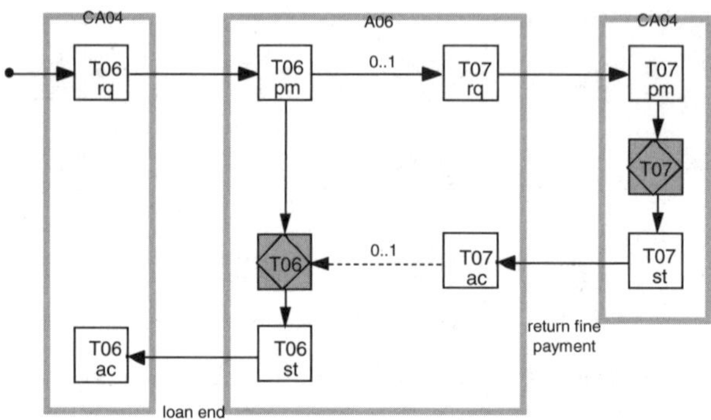

Fig. 7. Process Model of the loan end process

The corresponding, complete, part of the Action Model for actor A06 consists of the next action rules:

When T06(L) is requested, it must be declined if T05(L) is not accepted; otherwise it must be promised.
When T06(L) is promised, T07(L) must be requested if the acceptance of T05(L) was too late; otherwise T06(L) is executed and stated.
When T07(L) is stated, it must be rejected if the amount paid is not correct; otherwise it must be accepted.
When T07(L) is accepted, T06(L) is executed and stated.

Note that T07 is an optional enclosed transaction in T06. It will only be performed if applicable. Only in that case the last two rules are applied.

Fig. 8 exhibits the part of the State Model that corresponds with the two loan processes, as presented and discussed above, according to the diagrammatic language of WOSL [6], which is based on ORM [10]. It shows, among other things, the rule that a person may have more than one membership at the same time, and that a loan must

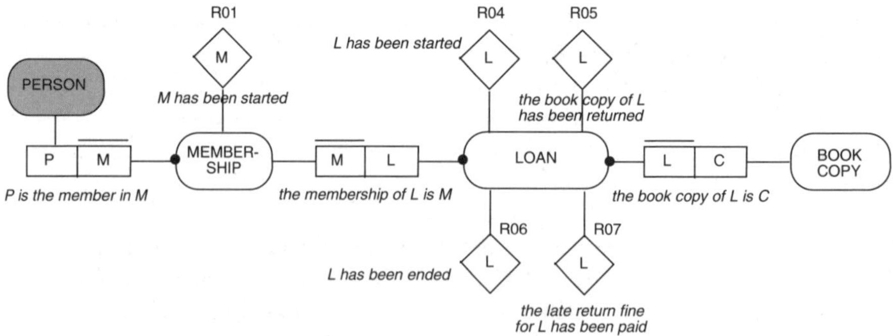

Fig. 8. State Model of (a part of) the Library

have an associated membership and an associated book copy. The diamond shaped unary fact types are transaction results. They are the only fact types that can be created by actors, together with the objects to which they belong. All other fact types are existentially dependent on them. The object class PERSON is colored gray to express that its members are created outside the scope of the Library, contrary to MEMBERSHIP, LOAN, and BOOK COPY.

Lastly, let us address the comments we have put to citations of the knowledge sources in Sec. 1. Regarding the concept customer, two questions were raised: is it basic and is it atomic? We don't consider it basic because it is a role of a person or an institute. In the latter case it would even also be not atomic. The atomicity of the concept of order is questionable because it normally is an aggregation of things. The concept of quantity back-ordered is not considered basic because it normally would be computed.

4 Conclusions

Based on the Ψ-theory (see Section 2), a clear distinction can be made between the function and the construction of an enterprise, respectively called its business and its organization. Another useful separation of concerns that can be made subsequently regards the distinction between three aspect organizations: the B-organization, the I-organization, and the D-organization. The ontological model of an enterprise is the (Ψ-theory-based) model of its B-organization. Among other things, it contains a complete set of business object classes, business fact types, business event types, and business rules.

A business object class is the extensional counterpart of an (intensional) unary fact type, and thus not a separate concept [17]. As an example, the class CAR = {x | car(x)}. Consequently, there is only one concept needed to describe the state of a (business) world, which is the concept of business fact.

A business event is the occurrence of a transition in the coordination world of an enterprise's B-organization. It is the effect of an act by an actor in the B-organization. Business events are agenda for actors, i.e., things to which they respond by taking appropriate actions.

A business rule is a statement that constrains either the state space or the transition space of either the production world or the coordination world of an enterprise's B-organization. Defined in this way, we speak of a declarative-shape rule. It appears very practical to transform declarative-shape rules into imperative-shape rules, i.e., action rules. Ross' projectors [14] seem to be action rules.

Although there is no fundamental difference between the declarative way of formulating rules and the imperative way, one could argue that imperative-shape rules offer less freedom to act than declarative-shape rules. It is good to realize, however, that this is only a matter of appearance. An interesting topic for future research would be the relationship between the kind of an organization and the preference for one of the two shapes of rules. Our hypothesis is that it is likely to find a preference for declarative-shape rules in organizations where people have a high level of education and professionalism. Conversely, one may expect to find a preference for imperative-shape rules in organizations where this level is low. Compliance with the rules is in the first kind of organizations a matter of trust in the competence and the responsibility of people. In the second kind, it is more likely that compliance is enforced by (automated) workflow systems, in which the rules are 'hard-wired'.

Next to business rules, so the rules applicable to the B-organization, there are similar operational rules concerning the I-organization and the D-organization. Examples of an I-rule and a D-rule are respectively:

Customers must be informed about penalties of late return before the car rental starts (i.e., before they accept the rental by signing the contract).
A copy of the driver license must be made before the customer fills out his data.

These I-rules and D-rules are certainly not unimportant. At the same time, it is obvious that the impact of violating them on the business is far less than the impact of violating B-rules. Therefore, it seems to be a good idea to deal with them separately. Derivation rules (Halpin [10]) or producers (Ross [14]) are infological rules; they belong to the I-organization. A subtle but important distinction can be made between the ontological definition of a fact and the infological rule by which it is computed or derived [6].

The analysis and discussion in this paper is performed in the context of Enterprise Engineering, where enterprises are considered to be designed artifacts. Business rules are part of the design and engineering of an enterprise, starting from its Enterprise Ontology. This design has been guided by the design principles of the applied Enterprise Architecture [8]. Business rules guide the operation of an enterprise; design principles guide its design.

References

1. Bunge, M.A.: Treatise on Basic Philosophy. The Furniture of the World, vol. 3. D. Reidel Publishing Company, Dordrecht (1977)
2. Bunge, M.A.: Treatise on Basic Philosophy. A World of Systems, vol. 4. D. Reidel Publishing Company, Dordrecht (1979)
3. Denning, P., Medina-Mora, R.: Completing the loops. In: ORSA/TIMS Interfaces, vol. 25, pp. 42–55 (1995)
4. Dietz, J.L.G.: Generic recurrent patterns in business processes. In: van der Aalst, W.M.P., ter Hofstede, A.H.M., Weske, M. (eds.) BPM 2003. LNCS, vol. 2678, Springer, Heidelberg (2003)
5. Dietz, J.L.G., Halpin, T.A.: Using DEMO and ORM in concert – A Case Study. In: Siau, K. (ed.) Advanced Topics in Database Research, vol. 3, IDEA Publishing, London (2004)
6. Dietz, J.L.G.: A World Ontology Specification Language. In: Meersman, R., Tari, Z. (eds.) Proceedings OTM 2005 Workshops. LNCS, vol. 3762, pp. 688–699. Springer, Heidelberg (2005)
7. Dietz, J.L.G.: Enterprise Ontology – Theory and Methodology. Springer, Heidelberg (2006)
8. Dietz, J.L.G., Hoogervorst, J.A.P.: Enterprise Ontology and Enterprise Architecture – how to let them evolve into effective complementary notions. GEAO Journal of Enterprise Architecture 2(1) (March 2007)
9. Gero, J.S., Kannengiesser, U.: The situated function-behaviour-structure framework. Design Studies 25(4), 373–391 (2004)

10. Halpin, T.A.: Information Modeling and Relational Databases. Morgan Kaufmann, San Francisco (2001)
11. Langefors, B.: Information System Theory. Information Systems 2, 207–219 (1977)
12. Morgan, T.: Business Rules and Information Systems. Addison-Wesley, Reading (2002)
13. OMG SVBR, http://www.omg.org/docs/bei/05-08-01.pdf
14. Ross, R.G.: Principles of the Business Rules Approach. Addison-Wesley, Reading (2003)
15. Searle, J.R.: The Construction of Social Reality, Allen Lane. The Penguin Press, London (1995)
16. Winograd, T., Flores, F.: Understanding Computers and Cognition: A New Foundation for Design, Ablex, Norwood (1986)
17. Wittgenstein, L.: Tractatus logico-philosophicus (German text with an English translation by C.K. Ogden). Routledge & Kegan Paul Ltd., London (1922)

Process Flexibility: A Survey of Contemporary Approaches

Helen Schonenberg, Ronny Mans, Nick Russell, Nataliya Mulyar, and Wil van der Aalst

Eindhoven University of Technology
P.O. Box 513, 5600 MB Eindhoven, The Netherlands
{m.h.schonenberg,r.s.mans,n.c.russell,nmulyar,w.m.p.v.d.aalst}@tue.nl

Abstract. Business processes provide a means of coordinating interactions between workers and organisations in a structured way. However the dynamic nature of the modern business environment means these processes are subject to a increasingly wide range of variations and must demonstrate flexible approaches to dealing with these variations if they are to remain viable. The challenge is to provide flexibility and offer process support at the same time. Many approaches have been proposed in literature and some of these approaches have been implemented in flexible workflow management systems. However, a comprehensive overview of the various approaches has been missing. In this paper, we take a deeper look into the various ways in which flexibility can be achieved and we propose an extensive taxonomy of flexibility. This taxonomy is subsequently used to evaluate a selection of systems and to discuss how the various forms of flexibility fit together.

Keywords: Taxonomy, flexible PAIS, design, change, deviation, underspecification.

1 Introduction

In order to retain their competitive advantage in today's dynamic marketplace, it is increasingly necessary for enterprises to streamline their processes so as to reduce costs and to improve performance. Moreover, it is clear that the economic success of an organisation is highly dependent on its ability to react to changes in its operating environment.

To this end, Process-Aware Information Systems (PAISs) are an desirable technology as these systems support the business operations of an enterprise based on models of both the organisation and its constituent processes. PAISs encompass a broad range of technologies ranging from systems which rigidly enforce adherence to the underlying process model, e.g., workflow systems or tracking systems, to systems which are guided by an implied process model but do nothing to ensure that it is actually enforced, e.g., groupware systems.

Typically, these systems utilise an idealised model of a process which may be overly simplistic or even undesirable from an operational standpoint. Furthermore the models on which they are based tend to be rigid in format and are not

J.L.G. Dietz et al. (Eds.): CIAO! 2008 and EOMAS 2008, LNBIP 10, pp. 16–30, 2008.

able to easily encompass either foreseen or unforeseen changes in the context or environment in which they operate. Up to now, there have not been any broadly adopted proposals or standards offering guidance for developing flexible process models able to deal with these sorts of changes. Instead most standards focus on a particular notation (e.g., XPDL, BPEL, BPMN, etc.) and these notations typically abstract from flexibility issues.

Process flexibility can be seen as the ability to deal with both foreseen and unforeseen changes, by varying or adapting those parts of the business process that are affected by them, whilst retaining the essential format of those parts that are not impacted by the variations. Or, in other words, flexibility is as much about what should stay the same in a process as what should be allowed to change[15]. Different kinds of flexibility are needed during the BPM life cycle of a process. Based on an extensive survey of literature and flexibility support offered by existing tools[1], a range of approaches to achieve process flexibility have been identified. These approaches have been described in the form of a taxonomy which provides a comprehensive catalogue of process flexibility approaches for the control-flow perspective.

The remainder of this paper is organised as follows. Section 2 presents the taxonomy for process flexibility. In Section 3, we use the taxonomy to evaluate the support of process flexibility in several contemporary PAISs, namely ADEPT1, YAWL, FLOWer and Declare. In Section 4 we discuss related work. Finally, we conclude the paper and identify opportunities for future work in Section 5.

2 Taxonomy of Flexibility

In this section, we present a comprehensive description of four distinct approaches that can be taken to facilitate flexibility within a process. All of these strategies improve the ability of business processes to respond to changes in their operating environment without necessitating a complete redesign of the underlying process model, however they differ in the timing and manner in which they are applied. Moreover they are intended to operate independently of each other. These approaches are presented in the form of a taxonomy which aims to define each of them in detail. The taxonomy is applicable to both classical (imperative) and constraint-based (declarative) specifications.

2.1 Specification Approaches

Generally, process behaviour depends on the structure of a process, which can be defined in an imperative or a declarative way. An *imperative approach* focuses on the precise definition of how a given set of tasks has to be performed (i.e., the task order is explicitly defined). In imperative languages, constraints on the execution order are described either via links (or connectors) between tasks and/or data conditions associated with them. A *declarative approach* focuses on *what* should be done instead of *how*. It uses constraints to restrict possible task execution

[1] See [10] for full details of the approach pursued and the literature and tools examined.

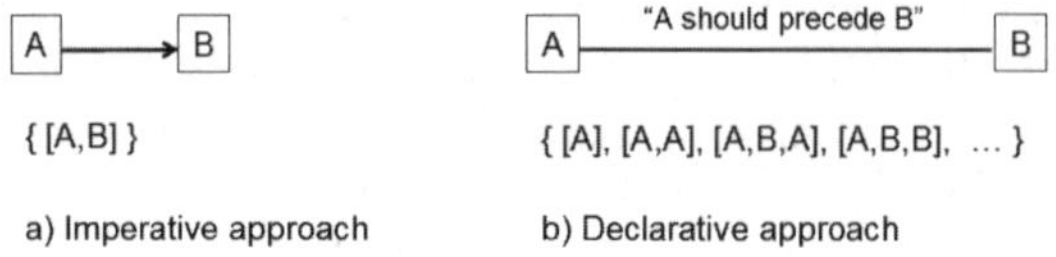

Fig. 1. Execution variances between imperative and declarative approaches

options. By default all execution paths are allowed, i.e., allowing all executions that do not violate the constraints. In general, the more constraints are defined for a process, the less execution paths are possible, i.e., constraints limit process flexibility. In declarative languages, constraints are defined as relations between tasks. Mandatory constraints are strictly enforced, while optional constraints can be violated, if needed. Figure 1 provides an example of both approaches. For both of them, a set of possible execution paths are illustrated. Note that a declarative approach offers many more execution paths.

2.2 Flexibility Types in Detail

In this section we discuss the individual flexibility types. Each of them is described in detail using a standard format including: a motivation, definition, scope, realisation options, i.e., the situations and domains to which the flexibility type applies, an example and discussion.

Flexibility by Design

Motivation. When a process operates in a dynamic environment it is desirable to incorporate support for the various execution alternatives that may arise within the process model. At runtime, the most appropriate execution path can be selected from those encoded in the design time process model.

Definition. *Flexibility by Design* is the ability to incorporate alternative execution paths within a process model at design time allowing the selection of the most appropriate execution path to be made at runtime for each process instance.

Scope. Flexibility by design applies to any process which may have more than one distinct execution trace.

Realisation options. The most common options for realisation of flexibility by design are listed below. It is not the intention of the authors to give a complete overview of all options.

- parallelism – the ability to execute a set of tasks in parallel;
- choice – the ability to select one or more tasks for subsequent execution from a set of available tasks;
- iteration – the ability to repeatedly execute a task[2];

[2] Note that iteration can be seen as a particular type of choice, where the join precedes the split.

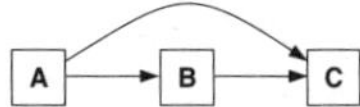

Fig. 2. Flexibility by design: a choice of execution paths is specified

- interleaving – the ability to execute each of a set of tasks in any order such that no tasks execute concurrently;
- multiple instances – the ability to execute multiple concurrent instances of a task; and
- cancellation – the ability to withdraw a task from execution now or at any time in the future.

The notions above are thoroughly described by the workflow patterns [20] and have been widely observed in a variety of imperative languages. We argue that these concepts are equally applicable in a declarative setting which has a much broader repertoire of constraints that allow for flexibility by design. Note that both approaches really differ with respect to flexibility. To increase flexibility in an imperative process, more execution paths have to be modeled explicitly, whereas increasing flexibility in declarative processes is accomplished by reducing the number of constraints, or weakening existing constraints.

Example. Figure 2 exemplifies a choice construct in an imperative model. The figure depicts that after executing A, it is possible to either execute B, followed by C, or to execute C directly. Using the choice construct, the notion of skipping tasks can be predefined in the process model.

Discussion. Realisation options can be implemented differently in different ways. For example there are different variants of the choice construct, such as exclusive choice and deferred choice, which can be effected in different ways. Interested readers are referred to the workflow patterns [20].

Describing all possible execution paths in a process model completely at design-time may be either undesirable from the standpoint of model complexity or impossible due to an unknown or unlimited number of possible execution paths. The following three flexibility types provide alternative mechanisms for process flexibility.

Flexibility by Deviation

Motivation. Some process instances need to temporarily deviate from the execution sequence described by the associated process model in order to accommodate changes in the operating environment encountered at runtime. For example, it may be appropriate to swap the ordering of the *register patient* and *perform triage* tasks in an emergency situation. The overall process model and its constituent tasks remain unchanged.

Definition. *Flexibility by Deviation* is the ability for a process instance to deviate at runtime from the execution path prescribed by the original process without altering its process model. The deviation can only encompass changes to the execution sequence of tasks in the process for a specific process instance, it does not allow for changes in the process model or the tasks that it comprises.

Scope. The concept of deviation is particularly suited to the specification of process models which are intended to guide possible sequences of execution rather than restrict the options that are available (i.e., they are descriptive rather than prescriptive). These specifications contain the preferred execution of the process, but other scenarios are also possible.

Realisation options. The manner in which deviation is achieved depends on the specification approach utilised. Deviation can be seen as varying the actual tasks that will be executed next, from those that are implied by the current set of enabled tasks in the process instance. In imperative languages this can be achieved by applying deviation operations. For declarative approaches, deviation basically occurs through violation of optional constraints. The following set of operations characterise support for deviation by imperative languages:

- Undo *task A*: Moving control to the moment before the execution of *task A*. One point to consider with this operation is that it does not imply that the actions of the task are undone or reversed. This may be an issue if the task uses and changes data elements during the course of its execution. In such situations, it may also be desirable to roll-back or compensate for the consequences of executing the task in some way, although it is not always possible to do so, e.g., the effects of sending a letter can not be reversed.
- Redo *task A*: Executing a disabled, but previously executed *task A* again, without moving control. This operation provides the ability to repeat a preceding task. One possible use for the operation is to allow incorrectly entered data during task execution to be entered again. For example after registering a patient in a hospital and undertaking some examinations, the registration task can be repeated to adjust outdated or incorrect data. Note that updating registration data should not require medical examinations to be performed again.
- Skip *task A*: Passing the point of control to a task subsequent to an enabled *task A*. There is no mechanism to compensate for the skipped task by executing it at a later stage of the execution. This operation is useful for situations, where a (knowledgeable) user decides that it is necessary to continue execution, even though some preceding actions have not been performed. For example, in life threatening situations it should be possible to start surgery immediately, whereas normally the patient's health status is evaluated before commencing surgery.
- Create additional instance of *task A*: Creating an additional instance of a task that will run in parallel with those instances created at the moment of task instantiation. It should be possible to limit the maximal number of task instances running in parallel. For example, a travel agency making trip arrangements for a group of people has to do the same arrangements if the number of travelling people increase (i.e., a separate reservation has to be done for each person).
- Invoke *task A*: Allows a task in the process model that is not currently enabled, and has not yet been executed, to be initiated. This task is initiated

Fig. 3. Flexibility by deviation: the point of control is moved

immediately. For example, when reviewing an insurance claim, it is suspected that the information given may be fraudulent. In order to determine how to proceed, the next task to be executed is deferred and a detailed investigation task (which normally occurs later in the process) is invoked. The execution of the investigation task does not affect the thread of control in the process instance and upon completion of the invoked task, execution continues from this point. Should the thread of control reach a previously invoked task at a later time in a process instance, it may be executed again or skipped on a discretionary basis.

Note that although we define deviation operations for imperative approaches only, this does not mean that there is no notion of these deviations in declarative approaches. Consider for example constraint *"A precedes B"*, which is defined as an optional constraint. By executing B before any occurrence of A, A is actually skipped by violating the optional precedence constraint. In this paper we clearly make a distinction between deviation for imperative and declarative approaches, due to the subtle difference in the act of deviating. Providing a full mapping of deviation operations to declarative constraints is beyond the scope of this paper.

Example. Figure 3 exemplifies flexibility by deviation by applying a skip operation. In Figure 3(a) task B is enabled. After applying *skip* B (Figure 3(b)), it is possible to execute a (currently not enabled) successor of an enabled task B.

Discussion. Deviation operations can be implemented in different ways, but for process mining purposes it should be possible to identify where deviations occurred during process execution. Furthermore additional requirements for the operators can be given, e.g., the *"undo A"* operation only has any effect when task A has been executed previously. When undoing task A, it may be recorded in one of two possible ways in the execution trace: either the undo task is explicitly marked as an execution action, or the occurrence of task A being undone is removed from the trace.

Flexibility by Underspecification

Motivation. When specifying a process model it might be foreseen that in the future, during run-time execution, more execution paths are needed which must be incorporated within the existing process model. Furthermore, often only during the execution of a process instance does it become clear what needs to be done at a specific point in the process. When all execution paths cannot be defined in advance, it is useful to be able to execute an incomplete process model and dynamically add process fragments expressing missing scenarios to it.

Definition. *Flexibility by Underspecification* is the ability to execute an incomplete process model at run-time, i.e., one which does not contain sufficient information to allow it to be executed to completion. Note that this type of flexibility does not require the model to be changed at run-time, instead the model needs to be completed by providing a concrete realisation for the undefined parts.

Scope. The concept of underspecification is mostly suitable for processes where it is clearly known in advance that the process model will have to be adjusted at *specific points*, although the exact content at this point is not yet known (and may not be known until the time that an instance of the process is executed). This approach to process design and enactment is particularly useful where distinct parts of an overall process are designed and controlled by different work groups, but the overall structure of the process is fixed. In this situation, it allows each of them to retain some degree of autonomy in regard to the tasks that are actually executed at runtime in their respective parts of the process, whilst still complying with the overall process model.

Realisation options. An incomplete process model is deemed to be one which is well-formed but does not have a detailed definition of the ultimate realisation of every task. An incomplete process model contains one or more so-called *placeholders*. Placeholders are nodes which are marked as *underspecified* (i.e., their content is unknown) and whose content is specified during the execution of these placeholders. We distinguish two types of *placeholder enactment*:

- *Late binding*: where the realisation of a placeholder is selected from a set of available process fragments. Note that to realise a placeholder one process fragment has to be selected from an existing set of predefined process fragments. This approach is limited to selection, and does not allow a new process fragment to be constructed.
- *Late modelling*: where a new process fragment is constructed in order to realise a given placeholder. Not only can a process fragment be constructed from a set of currently available process fragments, but also a new process fragment can be developed from scratch[3]. Therefore late binding is encompassed by late modelling. Some approaches [21] limit the construction of new models by (declarative) constraints.

For both approaches, the realisation of a placeholder can occur at a number of distinct times during process execution. Here, two distinct *moments for realisation* are recognised:

- *before placeholder execution*: the placeholder is realised at commencement of a process instance or during execution before the placeholder has been executed for the first time.
- *at placeholder execution*: the placeholder is realised when it is executed.

[3] However, this can only be done by highly skilled persons.

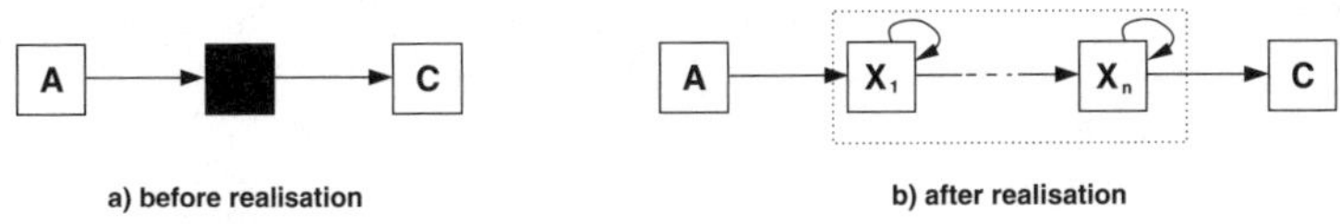

Fig. 4. Flexibility by underspecification: realisation of a placeholder

Placeholders can be either realised once, or for every subsequent execution of the placeholder. We distinguish two distinct *realisation types*:

- *static realisation*, where the process fragment chosen to realise the placeholder during the first execution is used to realise the placeholder for every subsequent execution.
- *dynamic realisation*, where the realisation of a placeholder can be chosen again for every subsequent execution of the placeholder.

Example. Figure 4(a) shows an incomplete process model with a placeholder task between A and C. Figure 4(b) illustrates the realisation of the placeholder, by a process fragment from a linked repository of process fragments. This figure shows the realisation as a sequence of self-looping tasks, but it can be realised by any well-formed process fragment.

Discussion. The process fragments available for placeholder realisation can be stored in a so called repository. A repository can be available for one or more processes, just for a particular task or a set of tasks.

Flexibility by Change

Motivation. In some cases, events may occur during process execution that were not foreseen during process design. Sometimes these events cannot be addressed by temporary deviations from the existing process model, but require the addition or removal of tasks or links from the process model on a permanent basis. This may necessitate changes to the process model for one or several process instances; or where the extent of the change is more significant, it may be necessary to change the process model for all currently executing instances.

Definition. *Flexibility by Change* is the ability to modify a process model at run-time such that one or all of the currently executing process instances are migrated to a new process model. Unlike the previously mentioned flexibility types the model constructed at design time is modified and one or more instances need to be transferred from the old to the new model.

Scope. Flexibility by change allows processes to adapt to changes that are identified in the operating environment. Changes may be introduced both at the level of the process instance and also at that of the process model (also known as change at instance level, and type or model level respectively).

Realisation options. For flexibility by change we distinguish the following variation points, which are partly based on [2].

Effect of change defines whether changes are performed on the level of a process instance or on the level of the process model, and what the impact of the change on the new process instances is.

- *Momentary change*: a change affecting the execution of one or more selected process instances. The change performed on a given process instance does not affect any future instances.
- *Evolutionary change*: a change caused by modification of the process model, affecting all new process instances.

Moment of allowed change specifies the moment at which changes can be introduced in a given process instance or a process model.

- *Entry time*: changes can be performed only at the moment the process instance is created. After the process instance has been created, no further changes can be introduced to the given process instance. Momentary changes performed at entry time affect only a given process instance. The result of evolutionary changes performed at entry time is that all new process instances have to be started after the change of the process model has been performed, and no existing process instances are affected (they continue execution according to the process model with which they are associated).
- *On-the-fly*: changes can be performed at any time during process execution. Momentary changes performed on-the-fly correspond to customisation of a given process instance during its execution. Evolutionary changes performed on-the-fly impact both existing and new process instances. The new process instances are created according to the new process model, while the existing process instances may need to migrate from the existing process model to the new process model.

Migration strategy defines what to do with running process instances that are impacted by an evolutionary change.

- *Forward recovery*: affected process instances are aborted.
- *Backward recovery*: affected process instances are aborted (compensated if necessary) and restarted.
- *Proceed*: changes introduced are ignored by the existing process instances. Existing process instances are handled the old way, and new process instances are handled the new way.
- *Transfer*: the existing process instances are transferred to a corresponding state in the new process model.

Example. In Figure 5(a) we show a process model that is changed into the process model depicted in Figure 5(b) by removing task B. The effect of this change is that instances of the new process model will skip task B permanently.

Discussion. A very detailed description of change operations can be found in [24]. The authors propose using high level change patterns rather than low level change primitives and give full descriptions for the identified patterns. Based on these change patterns and features, they provide a detailed analysis and evaluation of selected systems from both academia and industry.

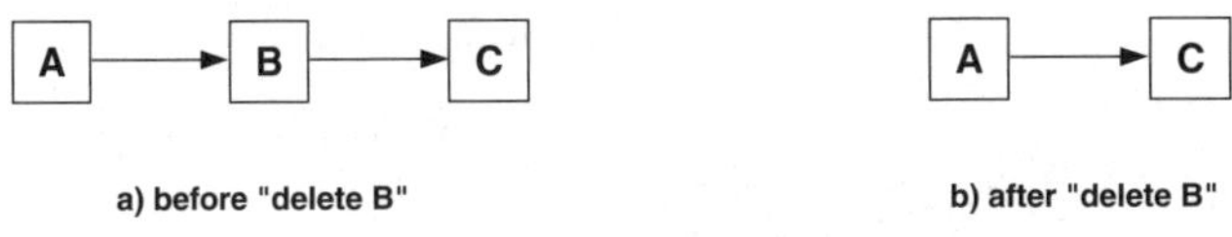

Fig. 5. Flexibility by change: a task is deleted

3 Evaluation of Contemporary Offerings

In this section, we apply the taxonomy presented in Section 2 to evaluate a selection of PAISs, namely, ADEPT1 [17], YAWL[4] (version 8.2b) [1,5,4], FLOWer (version 3.0) [3] and Declare (version 1.0) [11,12]. This evaluation provides an insight into the manner and extent to which the individual flexibility types are actually implemented in practice. The selection of PAISs has been based on the criterion of supporting process flexibility, which excludes classical workflow systems and most commercial systems. Moreover, the selected systems cover distinct areas of the PAIS technology spectrum, such as adaptive workflow (ADEPT1), case handling (FLOWer) and declarative workflow (Declare).

Although we focus on the evaluation of flexibility support, it is worthwhile mentioning that there is a huge difference in the maturity of the selected offerings. FLOWer has been widely used in industry where its flexible approach to case handling has proven to be extremely effective for a variety of purposes (e.g., insurance claim handling). ADEPT1 has also been successfully applied in different areas, such as health care [17]. Throughout its development lifetime, the designers of ADEPT [16] have focussed on supporting various forms of change [16,19,24]. The intention of the next version (ADEPT2) is to provide full support for changes, including transfer. YAWL is a more recent initiative based on formal foundations that shows significant promise in the support of a number of distinct flexibility approaches. Declare is the most recent of the offerings examined and its declarative basis provides a number of flexibility features. Interestingly, it supports transfer of existing process instances to the new process model. In the declarative setting, transfer is easily supported because it is not necessary to find a matching state in the new process for each instance [12].

The evaluation results are depicted in Table 1, which shows whether a system provides full (+), partial (+/–) or no support (–) for an evaluation criterion. For the full description of the evaluation criteria and the detailed evaluations for each of the offerings, we refer readers to the associated technical report [10]. The remainder of this section discusses the evaluation results.

Flexibility by design can be provided in several ways. Parallelism, choice and iteration are fully supported by all systems. Interleaving, multiple instances and cancellation are not supported by all systems, but they are all supported by YAWL and Declare, although in different ways. Due to the nature of declarative languages, the designer is encouraged to leave choices to users at run-time. *Flexibility by deviation* is similarly supported by both FLOWer and Declare

[4] The evaluation of YAWL includes the so-called Worklet Service.

Table 1. Product evaluations

	ADEPT1	YAWL	FLOWer	Declare
Flexibility by design				
Parallelism	+	+	+	+
Choice	+	+	+	+
Iteration	+	+	+	+
Interleaving	−	+	+/−	+
Multiple instances	−	+	+	+
Cancellation	−	+	−	+
Flexibility by deviation				
Deviation operations (imperative languages)				
Undo	−	−	+	
Redo	−	−	+	
Skip	−	−	+	
Create additional instance	−	−	+/−	
Invoke task	−	−	+	
Deviation operations (declarative languages)				
Violation of constraints				+
Flexibility by underspecification				
Late binding	−	+	−	−
Late modelling	−	+	−	−
Static, before placeholder	−	−	−	−
Dynamic, before placeholder	−	−	−	−
Static, at placeholder	−	−	−	−
Dynamic, at placeholder	−	+	−	−
Flexibility by change				
Effect of change				
Momentary change	+	−	−	+
Evolutionary change	−	+	−	+
Moment of allowed change				
Entry time	+	−	−	+
On–the–fly	+	+	−	+
Migration strategies for evolutionary change				
Forward recovery	−	+	−	−
Backward recovery	−	+	−	−
Proceed	−	−	−	+
Transfer	−	+	−	+

despite their distinct conceptual foundations. FLOWer achieves this by supporting almost all of the deviation operations, whereas Declare allows for violation of optional constraints. *Flexibility by underspecification* is only supported by YAWL (through its Worklet service). *Flexibility by change* is supported by ADEPT1, YAWL and Declare. ADEPT1 supports momentary change, which is allowed both at entry-time and on-the-fly. As mentioned earlier, the ADEPT developers have undertaken comprehensive research into the issue of dynamic process change and it will be interesting to see this incorporated in the next release (ADEPT2) when it becomes available. Evolutionary change is supported by YAWL and Declare, but unlike Declare, YAWL only supports changes to the process model. For this reason, YAWL does not support momentary change, entry time change and a proceed strategy. Declare supports changes for process instances and for the process model and offers proceed and transfer strategies. Transfer will be applied to those instances for which the execution trace does not violate the new process model, otherwise the proceed strategy will be applied, see [12] for details.

None of the evaluated systems provides the full range of flexibility alternatives. YAWL focusses on providing flexibility by design and underspecification, ADEPT1 on flexibility by change, FLOWer on flexibility by deviation and Declare provides flexibility in several different areas: design, deviation, and change.

4 Related Work

The need for process flexibility has long been recognised [8,18] in the workflow and process technology communities as a critical quality of effective business processes in order for organisations to adapt to changing business circumstances. It ensures that the "fit" between actual business processes and the technologies that support them are maintained in changing environments. The notion of flexibility is often viewed in terms of the ability of an organisation's processes and supporting technologies to adapt to these changes [22,7]. An alternate view advanced by Regev and Wegmann [13] is that flexibility should be considered from the opposite perspective i.e., in terms of what stays the same not what changes. Indeed, a process can only be considered to be flexible if it is possible to change it without needing to replace it completely [14]. Hence flexibility is effectively a balance between change and stability that ensures that the identity of the process is retained [13].

There have been a series of proposals for classifying flexibility, both in terms of the factors which motivate it and the ways in which it can be achieved within business processes. Snowdon et al. [22] identify three causal factors: type flexibility (arising from the diversity of information being handled), volume flexibility (arising from the amount of information types) and structural flexibility (arising from the need to operate in different ways). Soffer [23] differentiates between short-term flexibility, which involves a temporary deviation from the standard way of working, and long-term flexibility, which involves changes to the usual way of working. Kumar and Narasipuram [9] distinguish pre-designed flexibility which is anticipated by the designer and forms part of the process definition and just-in-time responsive flexibility which requires an "intelligent process manager" to deal with the variation as it arises at runtime. Carlsen et al. [6] identify a series of desirable flexibility features for workflow systems based on an examination of five workflow offerings using a quality evaluation framework. Heinl et al. [8] propose a classification scheme with distinct approaches – flexibility by selection, where a variety of alternative execution paths are designed into a process, and flexibility by adaption, where a workflow is "adapted" (i.e., modified) to meet with the new requirements. Two distinct approaches to achieving each of these approaches are recognised: flexibility by selection can be implemented either by advance modelling (before execution time) or late modelling (during execution time) where as flexibility by adaption can be handled either by type or instance adaption. Van der Aalst and Jablonski [2] adopt a similar strategy for supporting flexibility. Moreover they propose a scheme for classifying workflow changes in detail based on six criteria: (1) reason for change, (2) effect of change, (3) perspectives affected, (4) kind of change, (5) when are changes allowed and (6) what to do with existing process instances. Regev et al. [14] provide an initial attempt at defining a taxonomy of the concepts relevant to business process flexibility. This taxonomy has three orthogonal dimensions: the abstraction level of the change, the subject of the change and the properties of the change. Whilst it incorporates elements of the research initiatives discussed above, it is not comprehensive in form and does not describe the relationships that exist between these concepts or link them to possible realisation approaches.

The individual flexibility types discussed in this paper are informed by a multitude of research initiatives in the workflow and BPM fields. It is not possible to discuss these in detail in the confines of this paper, however there is a detailed literature review of the area in [10].

5 Conclusion

In this paper we have proposed a comprehensive taxonomy of flexibility approaches achieved based on an extensive survey of contemporary offerings and literature in the field. The four distinct flexibility types, that make up the proposed taxonomy, differ with respect to the moment and the manner in which both foreseen and unforeseen behaviour can be handled in a process. On the basis of this taxonomy, we have evaluated the support for flexibility provided by various commercial and research offerings (these were the only offerings that demonstrated any sort of flexibility features).

The evaluation process revealed varying approaches to flexibility support in the selected offerings. Moreover, individual offerings tended to exhibit a degree of specialisation in their approach to process flexibility. Such a strict specialisation limits the use of offerings in practice, since they are not capable of accommodating foreseen and unforeseen behaviours in processes during different phases of the BPM cycle. We hope that insights provided in this paper might trigger the enhancement of existing tools and/or development of new ones with a view for providing a greater support for flexibility.

A logical future step in researching the process flexibility is the establishment of a formal taxonomy/ontology for process flexibility, which would allow realisations of each of the flexibility types to be compared in an objective and language-independent way. Furthermore, another interesting line of research is to assess the extent of flexibility that current processes demand with a view to determining which of the flexibility approaches are of most use in practice.

In this paper we concentrated on the control-flow perspective of a business process, other perspectives addressing data, resources and applications used in a process are also subject to change. Thus, it would be worthwhile to extend the taxonomy in order to incorporate these perspectives. Additionally, there are some interesting process mining challenges presented by systems that support deviation or change operations, as in these offerings there is the potential for individual process instances to execute distinct process models.

References

1. van der Aalst, W.M.P., ter Hofstede, A.H.M.: YAWL: Yet Another Workflow Language. Information Systems 30(4), 245–275 (2005)
2. van der Aalst, W.M.P., Jablonski, S.: Dealing with Workflow Change: Identification of Issues and Solutions. International Journal of Computer Systems, Science, and Engineering 15(5), 267–276 (2000)
3. van der Aalst, W.M.P., Weske, M., Grünbauer, D.: Case Handling: A New Paradigm for Business Process Support. Data and Knowledge Engineering 53(2), 129–162 (2005)
4. Adams, M., ter Hofstede, A.H.M., van der Aalst, W.M.P., Edmond, D.: Dynamic, Extensible and Context-Aware Exception Handling for Workflows. In: Curbera, F., Leymann, F., Weske, M. (eds.) Proceedings of the OTM Conference on Cooperative information Systems (CoopIS 2007). LNCS, vol. 4803, pp. 95–112. Springer, Heidelberg (2007)
5. Adams, M., ter Hofstede, A.H.M., Edmond, D., van der Aalst, W.M.P.: Worklets: A Service-Oriented Implementation of Dynamic Flexibility in Workflows. In: Meersman, R., Tari, Z., et al. (eds.) Proceeding of the OTM Conference on Cooperative Information Systems (CoopIS 2006). LNCS, vol. 4275, pp. 291–308. Springer, Heidelberg (2006)
6. Carlsen, S., Krogstie, J., Sølvberg, A., Lindland, O.I.: Evaluating Flexible Workflow Systems. In: Proceedings of the Thirtieth Hawaii International Conference on System Sciences (HICSS-30), Maui, Hawaii, IEEE Computer Society Press, Los Alamitos (1997)
7. Daoudi, F., Nurcan, S.: A Benchmarking Framework for Methods to Design Flexible Business Processes. Software Process Improvement and Practice 12, 51–63 (2007)
8. Heinl, P., Horn, S., Jablonski, S., Neeb, J., Stein, K., Teschke, M.: A Comprehensive Approach to Flexibility in Workflow Management Systems. In: WACC 1999: Proceedings of the international joint conference on Work activities coordination and collaboration, pp. 79–88. ACM, New York (1999)
9. Kumar, K., Narasipuram, M.M.: Defining Requirements for Business Process Flexibility. In: Workshop on Business Process Modeling, Design and Support (BPMDS 2006), Proceedings of CAiSE 2006 Workshops, pp. 137–148 (2006)
10. Mulyar, N.A., Schonenberg, M.H., Mans, R.S., Russell, N.C., van der Aalst, W.M.P.: Towards a Taxonomy of Process Flexibility (Extended Version). BPM Center Report BPM-07-11, BPMcenter.org (2007)
11. Pesic, M., van der Aalst, W.M.P.: A Declarative Approach for Flexible Business Processes Management. In: Eder, J., Dustdar, S. (eds.) BPM Workshops 2006. LNCS, vol. 4103, pp. 169–180. Springer, Heidelberg (2006)
12. Pesic, M., Schonenberg, M.H., Sidorova, N., van der Aalst, W.M.P.: Constraint-Based Workflow Models: Change Made Easy. In: Curbera, F., Leymann, F., Weske, M. (eds.) Proceedings of the OTM Conference on Cooperative information Systems (CoopIS 2007). LNCS, vol. 4803, pp. 77–94. Springer, Heidelberg (2007)

13. Regev, G., Bider, I., Wegmann, A.: Defining Business Process Flexibility with the Help of Invariants. Software Process Improvement and Practice 12, 65–79 (2007)
14. Regev, G., Soffer, P., Schmidt, R.: Taxonomy of Flexibility in Business Processes. In: Proceedings of the 7th Workshop on Business Process Modelling, Development and Support (BPMDS 2006) (2006)
15. Regev, G., Wegmann, A.: A Regulation-Based View on Business Process and Supporting System Flexibility. In: Workshop on Business Process Modeling, Design and Support (BPMDS 2005), Proceedings of CAiSE 2005 Workshops, pp. 35–42 (2005)
16. Reichert, M., Dadam, P.: ADEPTflex: Supporting Dynamic Changes of Workflow without Loosing Control. Journal of Intelligent Information Systems 10(2), 93–129 (1998)
17. Reichert, M., Rinderle, S., Dadam, P.: ADEPT Workflow Management System. In: van der Aalst, W.M.P., ter Hofstede, A.H.M., Weske, M. (eds.) BPM 2003. LNCS, vol. 2678, Springer, Heidelberg (2003)
18. Reijers, H.A.: Workflow Flexibility: The Forlorn Promise. In: 15th IEEE International Workshops on Enabling Technologies: Infrastructures for Collaborative Enterprises (WETICE 2006), Manchester, United Kingdom, June 26-28, 2006, pp. 271–272. IEEE Computer Society, Los Alamitos (2006)
19. Rinderle, S., Reichert, M., Dadam, P.: Correctness Criteria For Dynamic Changes in Workflow Systems: A Survey. Data and Knowledge Engineering 50(1), 9–34 (2004)
20. Russell, N., ter Hofstede, A.H.M., van der Aalst, W.M.P., Mulyar, N.: Workflow Control-Flow Patterns: A Revised View. BPM Center Report BPM-06-29, BPM-center.org (2006)
21. Sadiq, S.W., Sadiq, W., Orlowska, M.E.: Pockets of Flexibility in Workflow Specification. In: Kunii, H.S., Jajodia, S., Sølvberg, A. (eds.) ER 2001. LNCS, vol. 2224, pp. 513–526. Springer, Heidelberg (2001)
22. Snowdon, R.A., Warboys, B.C., Greenwood, R.M., Holland, C.P., Kawalek, P.J., Shaw, D.R.: On the Architecture and Form of Flexible Process Support. Software Process Improvement and Practice 12, 21–34 (2007)
23. Soffer, P.: On the Notion of Flexibility in Business Processes. In: Workshop on Business Process Modeling, Design and Support (BPMDS 2005), Proceedings of CAiSE 2005 Workshops, pp. 35–42 (2005)
24. Weber, B., Rinderle, S.B., Reichert, M.U.: Change Support in Process-Aware Information Systems - A Pattern-Based Analysis. Technical Report Technical Report TR-CTIT-07-76, ISSN 1381-3625, Centre for Telematics and Information Technology, University of Twente, Enschede (2007), http://eprints.eemcs.utwente.nl/11331/

Subsuming the BPM Life Cycle in an Ontological Framework of Designing

Udo Kannengiesser

NICTA, Australian Technology Park, Bay 15 Locomotive Workshop
Eveleigh NSW 1430, Australia
udo.kannengiesser@nicta.com.au

Abstract. This paper proposes a framework to represent life-cycle activities performed in business process management (BPM). It is based on the function-behaviour-structure (FBS) ontology that represents all design entities uniformly, independently of the specific stages in their life cycle. The framework specifies a set of distinct activities that operate on the function, behaviour and structure of a business process, subsuming the different life-cycle stages within a single framework. This provides an explicit description of a number of BPM issues that are inadequately addressed in current life-cycle models. They include design-time analysis, flexibility of tasks and sub-processes, interaction between life-cycle stages, and the use of experience.

Keywords: BPM, BPM life cycle, FBS ontology.

1 Introduction

The notion of business process management (BPM) has emerged as a paradigm in organisational research and practice. It includes various techniques and tools that support business processes through all stages in their life cycle. Four stages are often proposed to compose the BPM life cycle [1, 2, 3], Figure 1:

1. Process Design: This stage includes modelling existing ("as-is") or future ("to-be") business processes.
2. Process Implementation: This stage provides and prepares the systems that are to carry out the business process. Systems can include both human operators and software.
3. Process Enactment: This stage realises the "actual", instantiated process using the models and configurations produced by the first two stages.
4. Process Evaluation: This stage monitors, analyses and validates the "actual" process and feeds the results back to the design stage.

The BPM life cycle suggests an iterative, continuous approach to managing business activities, aiming to enable adaptation to changes in the business environment through redesign of the process. It has been used as a framework for locating BPM research and as a vision and benchmark for BPM tool vendors.

J.L.G. Dietz et al. (Eds.): CIAO! 2008 and EOMAS 2008, LNBIP 10, pp. 31–45, 2008.

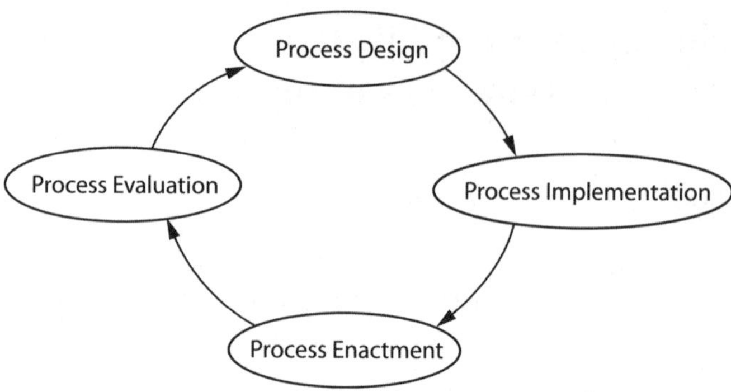

Fig. 1. The BPM life cycle

However, this model lacks explicit representations for a number of issues in BPM:

- Design-time analysis: The current model limits activities of analysis to the Process Evaluation stage. However, there is an increasing interest in approaches and tools for quantitative performance analysis, simulation and optimisation of business processes at design time [4].
- Local flexibility: Restricting all design capacity to the first life-cycle stage does not allow refining or adapting business processes closer to the Process Enactment stage. This makes business processes rigid and inaccessible for customisation, continuous improvement and control at the level of individual tasks or sub-processes.
- Interaction: The BPM life cycle suggests a sequential, top-down execution of the four stages, reminiscent of the "waterfall" model in software engineering. However, in practice the individual stages commonly overlap [1], and new process requirements often emerge during the life cycle leading to dynamic interactions between the stages.
- Use of experience: The individual stages do not show the role of experience that is gained from previous life-cycle activities to be reused in new life-cycle situations.

This paper proposes an ontological framework that captures the BPM life cycle with more explicit reference to the issues listed above. It is based on two fundamental ideas. Firstly, business processes are designed entities or artefacts that can be understood in the same way as physical artefacts such as houses, cars, computers, etc. Secondly, the life cycle of business processes can be subsumed in a uniform framework of designing. This expands the notion of designing to include activities that have traditionally been viewed outside its scope. The framework presented in this paper is based on the function-behaviour-structure (FBS) ontology [5, 6] that has been widely used in the broader field of design research.

2 An Ontological View of Business Processes

2.1 The Function-Behaviour-Structure Ontology

The FBS ontology distinguishes between three aspects of an artefact [5, 7]: function (F), behaviour (B) and structure (S). To provide a good understanding of this ontology (originally developed to represent physical products), Section 2 uses examples of artefacts from the domains of both (physical) products and business processes.

2.1.1 Function

Function (F) of an artefact is defined as its teleology ("what the artefact is for"). This definition views function as dependent on an observer's goals rather than on the artefact's embodiment. Function represents the usefulness of the artefact for another, "using system" [8]. It should not be confused with the concept of "transfer function".

Functions are often described using natural language expressions based on verb-noun pairs. For example, some of the functions of a window can be described as "to provide view", "to provide daylight" and "to provide rain protection".

Process goals (albeit defined in different ways by different people) represent an important class of functions of business processes. They replace particular states of the world (i.e., of the "using system") with ones that are more desirable from an individual point of view. For example, a function of the process "credit an account" could be formulated as the replacement of the state "not paid" with the state "paid".

Functions of a business process also comprise business goals such as "attract new customers" and "reduce time to market", and quality goals such as reliability and maintainability.

2.1.2 Behaviour

Behaviour (B) of an artefact is defined as the attributes that can be derived from its structure. They provide criteria for comparing and evaluating different objects or processes. In the window example, behaviours include "thermal conduction", "light transmission" and "direct solar gain". Typical behaviours of processes include speed, cost and accuracy. These behaviours can be specialised and/or quantified for instances of processes in particular domains. The notion of behaviour also covers "internal" attributes such as formal correctness and consistency.

2.1.3 Structure

Structure (S) of an artefact is defined as its components and their relationships. This definition can be understood most intuitively when applied to physical objects, such as windows, engines and buildings. Here, structure comprises an object's form (i.e., geometry and topology) and material. In the window example, form includes "glazing length" and "glazing height", and material includes "type of glass".

Mapping the concept of structure to business processes requires generalising the notions of form and material into macro-structure and micro-structure, respectively. Macro-structure is "formed" by the set of components and relationships that are distinguishable at a given level of abstraction. Micro-structure "materialises" macro-structure and is described using a shorthand qualifier, as its components and relationships are too fine-grained to be represented explicitly. In the window example, material is specified only as a label for the "type of glass" rather than as a set of molecular components and their relationships.

Three interconnected components form the macro-structure of a process, Figure 2: input, transformation, and output [6]. The transformation component often specifies a sequence of activities or states that are its sub-components (not shown in Figure 2). This description of process macro-structure maps onto Dietz' [8] "construction perspective" (or "white-box perspective"): The transformation component may be elementary ("black box" or "transfer function") or composite ("white box"), but the overall process structure can always be viewed as a "construction" of the same three components.

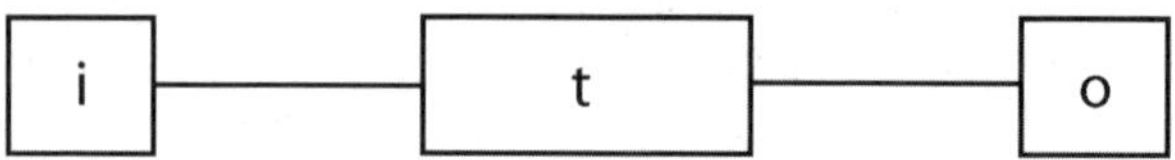

Fig. 2. Macro-structure of a process (i = input; t = transformation; o = output)

Two perspectives can be adopted to represent the micro-structure or "material" of a process:

1. Object-centred perspective: This perspective views micro-structure as the agent performing the transformation (t) and as the embodiment of the input (i) and output (o). The agent can be a specific person, department, organisation, software, role or a similar construct.
2. Process-centred perspective: This perspective views micro-structure as the underlying mechanism of the process. It can be understood as a "style of performing" the process, generalised from a set of more specific micro-activities. For example, a possible micro-structure of the business process "pay the supplier" may be labelled as "internet banking". The specific set of activities that "materialise" this business process, in terms of distinct steps such as "log in to online banking system", "fill out funds transfer form" and "click the submit button", are not shown. They are located at a clearly lower level of abstraction that is not of direct interest at the higher process level.

2.1.4 Relationships between Function, Behaviour and Structure

Humans construct relationships between function, behaviour and structure through experience and through the development of causal models based on interactions with the artefact. Function is ascribed to behaviour by establishing a teleological connection between the human's goals and measurable effects of the artefact. There is no direct relationship between function and structure [9]. Behaviour is derived from structure using physical laws or heuristics. This may require knowledge about external effects and their interaction with the artefact's structure. In the window example, deriving the behaviour "light transmission" requires considering external light sources. An example for processes is accuracy, which is a behaviour derived from the process output and an external benchmark.

2.2 FBS Views in the BPM Life Cycle

Different observers usually have different views of the function, behaviour and structure of an artefact. This Section shows how differences in FBS views of business processes are driven by different design concerns and life-cycle stages.

2.2.1 Views for Different Design Concerns

The concept of different views among designers is well-known, particularly in multi-disciplinary environments such as the architecture-engineering-construction (AEC) domain. In these environments, every discipline has a distinct set of design concerns that require the use of specific representations or views of the artefact. For example, an architectural view (i.e., one that addresses design concerns typical for architects) of the structure of a building design usually consists of a configuration of spaces, while a structural engineering view (i.e., one that addresses design concerns typical for structural engineers) of the same building usually consists of a configuration of floors and walls. The differences between these views of structure are based on the different behaviours that can be derived from them based on the functions that capture a specific design concern. Functions associated with the architectural view include spatial and environmental qualities, and functions associated with the structural engineering view include aspects of stability. Different views are also common within the same discipline; see, for instance, the "4+1" views of software architecture [10].

An example of orthogonal views in process modelling is the notion of different perspectives of a process, such as proposed by Curtis et al. [11]: the "task", the "control-flow", the "organisational" and the "informational" perspective.[1] A mapping of these perspectives onto the FBS ontology, Table 1, shows that they all relate to the notion of structure, including macro-structure (elementary and decomposed) and micro-structure (object- and some process-centred). The connection of the four perspectives to different design concerns has been pointed out by Luo and Tung [12].

2.2.2 Views for Different Life-Cycle Stages

Views of artefacts are further differentiated based on the life-cycle stages that deal with these artefacts. To capture both business processes and physical objects, a generic life cycle is specified comprising the following stages, Figure 3: Design, Implementation, Realisation, and Diagnosis. In the world of physical products, these stages are often known as Product Design, Production Planning, Production, and Product Testing. The remainder of this Section presents the FBS views typical for the individual stages, noting that these views are not always clear-cut.

FBS View in the Design Stage

For both physical products and business processes, the FBS view in Design comprises a model of structure that reflects the required functions and behaviours underpinning

[1] Curtis' original terms for the "task" and the "control-flow" perspective (namely the "functional" and the "behavioural" perspective, respectively) have not been adopted in this paper to avoid confusion with the notions of function and behaviour in the FBS ontology.

Table 1. Mapping four process perspectives onto the FBS ontology

Constructs in the FBS ontology	Process perspectives [11, p. 77]
i (elementary) t (elementary) o (elementary)	*Task*: "what process elements are being performed, and what flows of informational entities (e.g., data, artefacts, products), are relevant to these process elements"
t (decomposed into flows of activities)	*Control Flow*: "when process elements are performed (e.g., sequencing), as well as aspects of how they are performed through feedback loops, iteration, complex decision-making conditions, entry and exit criteria, and so forth"
object- and some process-centred micro-structure of i, t and o	*Organisational*: "where and by whom (which agents) in the organisation process elements are performed, the physical communication mechanisms used for transfer of entities, and the physical media and locations used for storing entities"
i (decomposed into information structures) t (decomposed into flows of information) o (decomposed into information structures)	*Informational*: "the informational entities produced or manipulated by a process; these entities include data, artefacts, products (intermediate and end), and objects; this perspective includes both the structure of informational entities and the relationships among them"

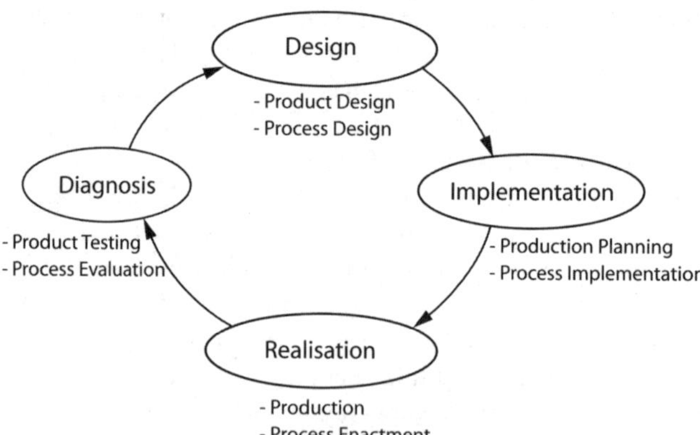

Fig. 3. A generic life cycle capturing both (physical) product life cycle and BPM life cycle

the design decisions in favour of that particular structure. This view may be partitioned according to specific design concerns, as outlined in Section 2.2.1. A number of languages have been developed to represent generic or concern-specific views of artefact structure in Design, often with tool support such as computer-aided drafting (CAD) packages for object models and BPM suites for business process models.

FBS View in the Implementation Stage

This view generates a model of the artefact that can readily be realised using the resources that are or can be made available. For mechanical assemblies, for example, this involves creating a set of manufacturing and assembly plans based on the object drawings received from the Design stage. These plans are procedural descriptions of the steps required to transform raw material into parts (specified in manufacturing plans) and parts into assemblies (specified in assembly plans). Plans are prepared in a way to be understood by human workers and/or by numerically controlled (NC) machines. The functions and behaviours associated with the implementation view predominantly deal with issues specific to artefact realisation, such as feasibility, production time and production cost.

For business processes, the FBS view in Implementation produces models of process structure in the form of enactment plans that can be understood by human process workers or automated process enactment systems. These models are often referred to as workflows. They include more details of process structure than is captured in the business process models of the design view. For example, workflows usually include process states such as "started" and "completed", to manage the orchestration of individual activities based on the resources available in the enactment (realisation) environment. Functions and behaviours of workflows concentrate on feasibility (including correctness, absence of deadlocks, etc.), time, resource utilisation and similar aspects, rather than the more high-level business goals reflected in business process models.

FBS View in the Realisation Stage

The FBS view in Realisation is identical to the FBS view in the Implementation stage, even though the structure of the realised artefact is no longer embodied in a representation medium (such as paper or computational media) but in the "real" world. Behaviours can be derived from this structure that can then be compared with the behaviours of the implemented (i.e., not yet realised) artefact. This is then a basis for devising control actions for process instances that deviate from the workflow.

FBS View in the Diagnosis Stage

The FBS view in Diagnosis is identical to the FBS view that was adopted in the Design stage. This allows evaluating the artefact by comparing the measured behaviour with the specified behaviour. As a possible result, improvements can be initiated by returning to the Design stage and thus commencing another life cycle.

Summary

In the generic life cycle, at its current level of granularity, there are only two different FBS views, each of which includes a design-time and a runtime component corresponding to distinct life-cycle stages:

- The *concept view*, adopted in the Design stage (design-time component) and the Diagnosis stage (runtime component)
- The *realisation view*, adopted in the Implementation stage (design-time component) and the Realisation stage (runtime component)

These views form the basis for integrating the entire life cycle of an artefact in a single framework of designing.

3 The BPM Life Cycle in a Framework of Designing

3.1 An Initial Framework of Designing

Designing aims to create the structure of new artefacts to meet a set of requirements stated as functions. As this mapping between function and structure can be established only via behaviour, that behaviour must satisfy two constraints: First, it must reliably describe the object's "actual" performance under operating conditions, and, second, it must be consistent with the functions required. One can think of behaviour as being located in a field of tension between desirability, represented by function, and feasibility, represented by structure. Designed objects are successful only if their desired behaviour (constrained by function) matches their feasible behaviour (constrained by structure).

Based on these concepts, an initial process framework of designing can be formulated comprising the following fundamental design activities [7]:

- *Formulation*: transforms required function into behaviour that is expected to achieve that function.
- *Synthesis*: transforms expected behaviour into a structure that is a candidate solution to the design problem.
- *Analysis*: transforms the structure of the candidate design solution into "actual" behaviour.
- *Evaluation*: compares expected behaviour and "actual" behaviour.
- *Documentation*: produces a description of the final design solution, in sufficient detail to carry out the next stage in the life cycle (i.e., implementation or realisation).
- *Reformulation*: modifies some of the properties of the artefact, affecting function, behaviour or structure.

This framework can be applied to any artefact represented in any FBS view, and thus also captures the two FBS views derived in Section 2.2.2. This results in two distinct design processes, *concept designing* and *realisation designing*, represented using the same framework. Note that the activity of analysis for these design processes covers both design-time and runtime analyses, based on the embodiment of the artefact in the "real" world or in a representation medium. Examples of design-time analyses of (represented) business processes include process simulation, verification and informal diagrammatic analysis [4]. An example of runtime analysis of ("real") business processes is business activity monitoring (BAM).

Concept designing subsumes the Design stage in the BPM life cycle, but extends this notion in two ways. First, it provides a more detailed account of designing as a set of distinct activities rather than as a black box. Second, it spans the traditional divide between the modelling and the operating environment, tying them more closely together and thus enabling responsiveness of designing to both design-time and runtime analyses.

Understanding realisation as designing accounts for the need to provide human operators with sufficient freedom for carrying out processes in a way adapted to the situation at hand [13]. Realisation designing generates two entities: one is an elaboration of the artefact, i.e. of the business process, and the other one is the set of activities to be performed for providing and preparing the systems that are to execute that process. The latter can be viewed as a separate, secondary artefact generated during realisation designing.

3.2 A Model of Three Interacting Worlds

The initial framework of designing presented in Section 3.1 is a basis for capturing the life-cycle aspects of design-time analysis and local flexibility; however, it does not address interaction and the use of experience. This Section introduces the foundations for an extended framework, drawing on a cognitively-based model of designing.

Designers perform actions in order to change their environment. By observing and interpreting the results of their actions, they then decide on new actions to be executed on the environment. The designers' concepts may change according to what they are "seeing", which itself is a function of what they have done. One may speak of an "interaction of making and seeing" [14]. This interaction between the designer and the environment strongly determines the course of designing. This idea is called situatedness, whose foundational concepts go back to the work of Dewey [15] and Bartlett [16].

Gero and Kannengiesser [5] have modelled situatedness using the idea of three interacting worlds: the external world, interpreted world and expected world, Figure 4(a).

The *external world* is the world that is composed of things outside the designer or design agent. No matter whether things are "real" or represented, we refer to all of them as just "design representations". This is because, in our model, their purpose is to support interpretation and communication of design agents.

The *interpreted world* is the world that is built up inside the design agent in terms of sensory experiences, percepts and concepts. It is the internal representation of that part of the external world that the design agent interacts with. The interpreted world provides an environment for analytic activities and discovery during designing.

The *expected world* is the world imagined actions of the design agent will produce. It is the environment in which the effects of actions are predicted according to current goals and interpretations of the current state of the world.

These three worlds are interrelated by three classes of interaction. *Interpretation* transforms variables that are sensed in the external world into sensory experiences, percepts and concepts that compose the interpreted world. *Focussing* takes some aspects of the interpreted world and uses them as goals for the expected world. *Action* is an effect which brings about a change in the external world according to the goals in the expected world.

Figure 4(b) presents a specialised form of this model, with the design agent (described by the interpreted and expected world) located within the external world, and with general classes of design representations placed into this nested model. The set of expected design representations (Xe^i) corresponds to the notion of a design state space, i.e. the state space of all possible designs that satisfy the set of requirements. This state space can be modified during the process of designing by transferring new interpreted design representations (X^i) into the expected world and/or transferring some of the expected design representations (Xe^i) out of the expected world. This leads to changes in external design representations (X^e), which may then be used as a basis for re-interpretation changing the interpreted world. (Changes in the external world may also

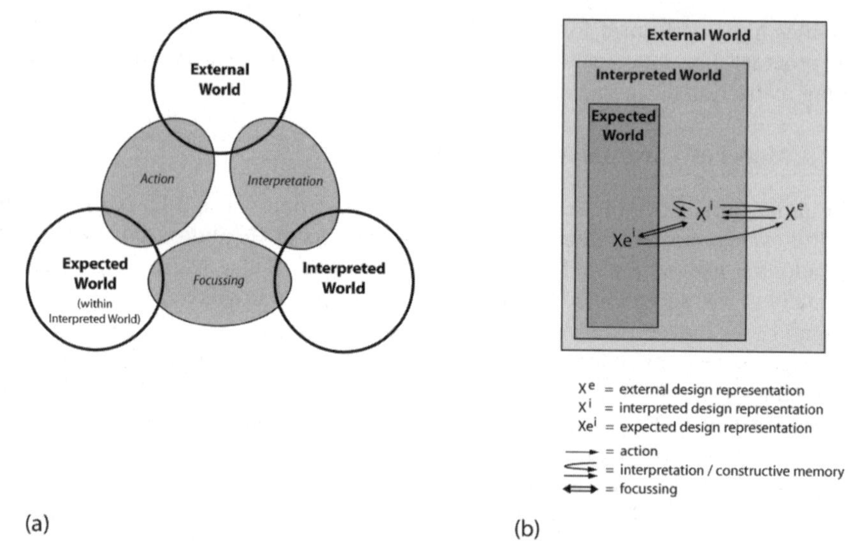

Fig. 4. Situatedness as the interaction of three worlds: (a) general model, (b) specialised model for design representations

occur independently of the design agent.) Novel interpreted design representations (X^i) may also be the result of memory (here called *constructive memory*), which can be viewed as a process of interaction among design representations within the interpreted world rather than across the interpreted and the external world.

Both interpretation and constructive memory are modelled as "push-pull" processes, i.e. the results of these processes are driven both by the original experience ("push") and by some of the agent's current interpretations and expectations ("pull") [17]. This notion captures the subjective nature of interpretation and constructive memory, using first-person knowledge grounded in the designer's interactions with their environment [17, 18, 19]. It is this subjectiveness that produces different views of the same entity. Note that the views presented in Section 2.2 are based on the generalised experience of disciplines and life-cycle concerns. Individuals construct views on the fly, emerging from the interplay of "push" and "pull" that potentially lead to novel interpretations over time.

3.3 Business Process Design in the Situated FBS Framework

Gero and Kannengiesser [5] have combined the FBS ontology with the model of interacting design worlds, by specialising the description of situatedness shown in Figure 4(b). In particular, the variable X, which stands for design representations in general, is replaced with the more specific representations F, B and S. This results in the situated FBS framework, Figure 5 [5]. In addition to using external, interpreted and expected F, B and S, this framework uses explicit representations of external requirements given to the designer by a stakeholder. Specifically, there may be external requirements on function (FR^e), behaviour (BR^e) and structure (SR^e). The situated FBS framework also includes the process of comparison between interpreted behaviour (B^i) and expected behaviour (Be^i), and a number of processes that transform interpreted structure (S^i) into interpreted behaviour (B^i), interpreted behaviour (B^i) into interpreted function (F^i), expected function (Fe^i) into expected behaviour (Be^i), and expected behaviour (Be^i) into expected structure (Se^i). Figure 5 uses the numerals 1 to 20 to label the resultant set of processes. They do not represent any order of execution. The 20 processes elaborate the fundamental design activities introduced in Section 3.1, which will be shown in the remainder of this Section.

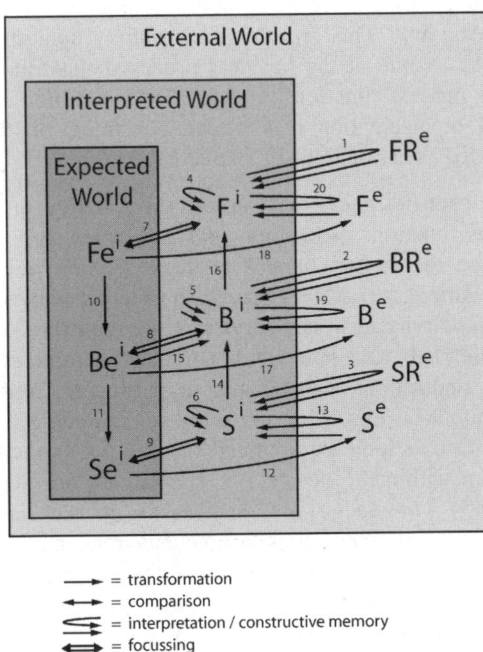

Fig. 5. The situated FBS framework (after [5])

3.3.1 Concept Designing of Business Processes

Formulation: Concept designers often receive an initial set of business and quality goals as FR^e, specific performance targets as BR^e, and some required (sequences of) activities as SR^e. The designers interpret these requirements (processes 1 – 3) and augment them by constructing additional requirements (processes 4 – 6). These are often requirements that relate to rather "common-sense" considerations, such as basic safety functions and reasonable throughput times. Concept designers ultimately decide on a subset of the requirements and concepts to be taken into consideration for generating design solutions (processes 7 – 9). A set of behaviours is derived from the functions considered (process 10).

Synthesis: Concept designers generate the structure of a business process that is expected to meet the required behaviours (process 11), and externalise that structure for communicating and/or reflecting on it (process 12). This is commonly done using standard notations such as BPMN, with appropriate tool support.

Analysis: Concept designers (or specialised analysis tools) interpret the externalised business process structure (process 13) and derive "actual" behaviours to allow for evaluation of the business process (process 14). If the external structure is "real" (i.e., executed), this activity corresponds to the Diagnosis stage of the life cycle.

Evaluation: consists of a comparison of expected behaviour and behaviour derived through analysis (process 15).

Documentation: When the evaluated business process design is satisfactory, concept designers produce an external representation of the final business process to be passed on to realisation designing. This representation mainly consists of business process structure (process 12), some of the business process behaviour (process 17) and, in few cases, business process function (process 18). A common example of behaviour in the externalised representation is a timing constraint on the business process. Functions are included as annotations in textual form.

Reformulation: Concept designers may, at any time, before or after documentation, focus on different function, behaviour and structure (processes 7 – 9). This reformulation can be driven by changes in the external requirements provided by stakeholders. For example, a customer may wish to increase the degree of automation by implementing an activity as a web service rather than through manual processing as they initially intended (i.e., new SR^e). Another example is a new requirement received from the realisation designer that a particular timing constraint of the business process cannot be met using the resources available (i.e., new BR^e). Other drivers of reformulation include requirements that are not explicitly stated as such but are constructed from within the designer. Examples include requirements emerging from (the designer's knowledge of) changes in market competition and new government regulations. Another common precursor for emerging design concepts is the detection of unsatisfactory behaviour through design-time or runtime analysis.

3.3.2 Realisation Designing of Business Processes

Formulation: Realisation designers usually receive little explicit requirements on function (FR^e) and behaviour (BR^e), as the documentation received from concept designing – the business process model – mostly represents required structure (SR^e). As a result, the interpretation of external requirements (processes 1 – 3) needs to be complemented through the internal construction of additional requirements (processes 4 – 6). Typically, the internally generated requirements are functions and behaviours related to a correct and resource-efficient orchestration of the given business process. Realisation designers can also construct elaborations of the process structure, for example, by dynamically allocating resources to activities. However, most modelling languages and practices in concept designing tend to over-specify process structure and thus restrict flexibility in process realisation [20, 21]. Realisation designers decide on the requirements and concepts to be considered in their workflow design (processes 7 – 9), and derive an additional set of behaviours from the functions considered (process 10).

Synthesis: Realisation designers generate a workflow structure that is expected to meet the required behaviours (process 11), and externalise that structure for communicating and/or reflecting on it (process 12). The structure may be expressed using diagrammatic notations to be understood by humans or using formal notations to be understood by automated systems for subsequent execution.

Analysis: Realisation designers (or specialised analysis tools) interpret the structure of the externalised workflow structure (process 13) and derive "actual" behaviours to allow for evaluation of the workflow (process 14). The external structure may be "real" (i.e., executed) or represented/simulated.

Evaluation: consists of a comparison of expected behaviour and behaviour derived through analysis (process 15).

Documentation: When realisation designers are satisfied with their evaluations, they produce an external representation of the final workflow. If the executing system is automated, this representation may include only process structure (process 12) and some behaviour (process 17). If the executing system involves a human process operator, some process function (process 18) can be added to facilitate understanding and acceptance of the workflow.

Reformulation: Realisation designers may, at any time, before or after business process execution, focus on different function, behaviour or structure (processes 7 – 9). This reformulation can be driven by changes in the external requirements provided by concept designers, usually in form of a modified structure of the higher-level business process model. This often occurs as a result of a reformulation of the concept design, as outlined in Section 3.3.1. Another common driver of reformulation is the detection of incorrect or inefficient activity execution (process 14), which may be addressed by performing local adaptations on a process instance level. In cases where this is not possible, the realisation designer needs to communicate the problem (process 17) and/or propose changes to the business process model (process 12). This communication leads to new requirements for the concept designer who then decides whether or not to take them into consideration.

4 Conclusion

Representing the BPM life cycle in the situated FBS framework provides a rich description of BPM activities, capturing the issues mentioned in Section 1:

- Design-time analysis: The framework comprises both design-time and runtime analyses. This is based on the uniform representation it provides for any type of embodiment of a business process, including paper-based, digital, simulated and "real" environments. No matter how the process is embodied, it can always be interpreted as structure that can then be transformed into behaviour.
- Local flexibility: Extending the scope of designing to include implementation and realisation provides flexibility at all levels of the life cycle, using the capability of change that is inherent to designing. All changes are based on the designers' decisions based on the current situation, constrained by their interpretation of the requirements.

- Interaction: Business processes can change at any time during their life cycle. The situated FBS framework captures this change through reformulation processes that operate on the function, behaviour or structure of a business process. Processes that have been reformulated in concept designing can be interpreted as new external requirements for realisation designing, and vice versa. This enables dynamic interactions between life-cycle stages.
- Use of experience: Experience is captured in the situated FBS framework by processes of interpretation and constructive memory. It is based not only on the currently active BPM life cycle but also on the designers' "life cycle", i.e., it is constructed from all their previous interactions with business process designs and with one another.

The explicit description of these issues can lead to a more profound understanding of the BPM life cycle. The design-ontological foundations established in this paper provide a tool for BPM research to gain access to a wider range of approaches drawn from various fields of design. We are currently using the situated FBS framework to address some of the major life-cycle related challenges faced by BPM practitioners [22], focussing on enhanced modelling languages and methods for more process flexibility, interoperability and alignment with business goals.

Acknowledgments. NICTA is a national research institute with a charter to build Australia's pre-eminent Centre of Excellence for information and communications technology (ICT). NICTA is building capabilities in ICT research, research training and commercialisation in the ICT sector for the generation of national benefit. NICTA is funded by the Australian Government as represented by the Department of Broadband, Communications and the Digital Economy and the Australian Research Council through the ICT Centre of Excellence program.

References

1. van der Aalst, W.M.P.: Business Process Management Demystified: A Tutorial on Models, Systems and Standards for Workflow Management. In: Desel, J., Reisig, W., Rozenberg, G. (eds.) ACPN 2003. LNCS, vol. 3098, pp. 1–65. Springer, Heidelberg (2004)
2. zur Muehlen, M., Ho, D.T.-Y.: Risk Management in the BPM Lifecycle. In: Bussler, C.J., Haller, A. (eds.) BPM 2005. LNCS, vol. 3812, pp. 454–466. Springer, Heidelberg (2006)
3. Wetzstein, B., Ma, Z., Filipowska, M., Bhiri, S., Losada, S., Lopez-Cobo, J.-M., Cicurel, L.: Semantic Business Process Management: A Lifecycle Based Requirements Analysis. In: Hepp, M., Hinkelmann, K., Karagiannis, D., Klein, R., Stojanovic, N. (eds.) Semantic Business Process and Product Lifecycle Management. Proceedings of the Workshop SBPM 2007, Innsbruck, Austria, pp. 1–10 (2007)
4. Vergidis, K., Tiwari, A., Majeed, B.: Business Process Analysis and Optimization: Beyond Reengineering. IEEE Transactions on Systems, Man, and Cybernetics – Part C: Applications and Reviews 38(1), 69–82 (2008)
5. Gero, J.S., Kannengiesser, U.: The Situated Function-Behaviour-Structure Framework. Design Studies 25(4), 373–391 (2004)

6. Gero, J.S., Kannengiesser, U.: A Function-Behavior-Structure Ontology of Processes. Artificial Intelligence for Engineering Design, Analysis and Manufacturing 21(4), 379–391 (2007)
7. Gero, J.S.: Design Prototypes: A Knowledge Representation Schema for Design. AI Magazine 11(4), 26–36 (1990)
8. Dietz, J.L.G.: Enterprise Ontology: Theory and Methodology. Springer, Berlin (2006)
9. de Kleer, J., Brown, J.S.: A Qualitative Physics Based on Confluences. Artificial Intelligence 24, 7–83 (1984)
10. Kruchten, P.: Architectural Blueprints – The "4+1" View Model of Software Architecture. IEEE Software 12(6), 42–50 (1995)
11. Curtis, B., Kellner, M.I., Over, J.: Process Modeling. Communications of the ACM 35(9), 75–90 (1992)
12. Luo, W., Tung, Y.A.: A Framework for Selecting Business Process Modeling Methods. Industrial Management & Data Systems 99(7), 312–319 (1999)
13. van Aken, J.E.: Design Science and Organization Development Interventions: Aligning Business and Humanistic Values. Journal of Applied Behavioral Science 43(1), 67–88 (2007)
14. Schön, D.A., Wiggins, G.: Kinds of Seeing and their Functions in Designing. Design Studies 13(2), 135–156 (1992)
15. Dewey, J.: The Reflex Arc Concept in Psychology. Psychological Review 3, 357–370 (1896 reprinted in 1981)
16. Bartlett, F.C.: Remembering: A Study in Experimental and Social Psychology. Cambridge University Press, Cambridge (1932 reprinted in 1977)
17. Smith, G.J., Gero, J.S.: What Does an Artificial Design Agent Mean by Being 'Situated'? Design Studies 26(5), 535–561 (2005)
18. Bickhard, M.H., Campbell, R.L.: Topologies of Learning. New Ideas in Psychology 14(2), 111–156 (1996)
19. Clancey, W.J.: Situated Cognition: On Human Knowledge and Computer Representations. Cambridge University Press, Cambridge (1997)
20. Goedertier, S., Vanthienen, J.: Declarative Process Modeling with Business Vocabulary and Business Rules. In: Meersman, R., Tari, Z., Herrero, P. (eds.) OTM-WS 2007, Part I. LNCS, vol. 4805, pp. 603–612. Springer, Heidelberg (2007)
21. Zhu, L., Osterweil, L.J., Staples, M., Kannengiesser, U., Simidchieva, B.I.: Desiderata for Languages to Be Used in the Definition of Reference Business Processes. International Journal of Software and Informatics 1(1), 37–65 (2007)
22. Bandara, W., Indulska, M., Chons, S., Sadiq, S.: Major Issues in Business Process Management: An Expert Perspective. BPTrends, 1–8 (October 2007)

Information Gathering for Semantic Service Discovery and Composition in Business Process Modeling

Norman May and Ingo Weber

SAP Research, Karlsruhe, Germany
{norman.may,ingo.weber}@sap.com

Abstract. When creating an execution-level process model today, two crucial problems are how to find the right services (service discovery and composition), and how to make sure they are in the right order (semantic process validation). While isolated solutions for both problems exist, a unified approach has not yet been available. Our approach resolves this shortcoming by gathering all existing information in the process, thus making the basis of semantic service discovery and task composition both broader and more targeted. Thereby we achieve the following benefits: (i) less modeling overhead for semantic annotations to the process, (ii) more information regarding the applicability of services, and (iii) early avoidance of inconsistencies in the interrelation between all process parts. Consequently, new or changed business processes can be realized in IT more efficiently and with fewer errors, thus making enterprises more agile in response to new requirements and opportunities.[1]

Keywords: Semantic Business Process Management, Web Service Discovery & Composition.

1 Introduction

When enterprises need to adapt to changes in their environment they often have to adjust their internal business processes. In such cases, a business expert with little technical knowledge designs a changed or new business process model, which should then be implemented. Today, the refinement of this model into an executable process is carried out manually, requiring time and communication between the domain expert and the respective IT expert with technical expertise. Hence, enterprises are confronted with limited adaptability of IT-supported business processes, leading to low agility and high cost for changes [3,18].

In this paper we present an improvement addressing this problem by extending and combining previous work on service discovery, service composition, and process validation. In our approach we use all available process information to

[1] This work has been funded through the German Federal Ministry of Economy and Technology (01MQ07012) and the European Union (IST FP6-026850, http://www.ip-super.org).

J.L.G. Dietz et al. (Eds.): CIAO! 2008 and EOMAS 2008, LNBIP 10, pp. 46–60, 2008.

reduce the modeling effort, to improve the efficiency of the refinement into an executable process, and to provide richer feedback on inconsistencies in models.

Our solution relies on semantic technologies which promise a simpler, cheaper and more reliable transformation from the business level into the technical level [3]. On the business level, the domain or business expert models a process in a graphic modeling notation, e.g. BPMN or UML activity diagrams, or in a process description language, e.g. a graphical representation of BPEL, and all tasks in the process are annotated with concepts of a common ontology.

Assuming that all tasks of the business process can be implemented by a set of Semantic Web services[2], an automatic translation of the business process into an orchestration of Semantic Web Services is possible. Efficient algorithms for such a translation require the Web services to be annotated with the same ontology as the business process [1,4,16]; i.e. every Web Service is annotated with its preconditions and its effect, i.e. the postcondition it establishes.

Finding the implementation for a process task is done in the following way: First, service discovery attempts to find a matching Web service. If this is not successful (or for other reasons not satisfactory, e.g., too costly), service composition is performed (e.g., [4]). Later on, the resulting combination of discovered or composed services and the process model needs to be validated to assure that the preconditions of all services hold in any possible case and that parallel services are not in conflict with one another. This is needed since in previous work discovery and composition is performed for every task in isolation and hence potential conflicts between multiple tasks are not taken into account directly.

Currently, modeling of a complex process is laborious due to the need to annotate all tasks with preconditions and postconditions manually. Also, by regarding the manually annotated precondition of a task, service composition may miss attractive solutions that could be applied by regarding the global process context (but not from the local task context). In addition, the supposedly "intelligent" tool may suggest faulty solutions, thus jeopardizing user acceptance.

Our solution is based on logical state summaries that may be encountered during any valid execution of the process. These states can, e.g., model real world behavior by using terms from the common ontology. The executable process is in an invalid state if the execution states of two parallel execution paths contradict each other. Such a contradiction arises in two cases: (1) One activity invalidates the precondition of another activity that is executed in parallel. (2) One activity relies on facts, some of which may be not true. We extend state-of-the-art approaches by integrating service composition and process validation to distribute information about the process model across all activities. More precisely, the contributions of this paper are the following:

1. Our method reduces the effort for modeling semantic business processes: The burden for annotating tasks is reduced because all context information is taken

[2] In this paper we focus on the description based on Semantic Web services for the sake of readability. However, any kind of known execution-level process building blocks which are described in an analogous same way may be used here, e.g., automated workflow tasks, or standardized but entirely manual activities.

into account. This is done, e.g. by expanding preconditions of the activities in the business process based on the postconditions of their predecessors.
2. During service composition our approach takes into account constraint-sets which restrict the states considered as intermediate states during service composition. The pruned states would otherwise lead to inconsistencies with states computed in parallel tasks.
3. We perform process validation in parallel with service composition. As a consequence, we are able to detect semantically invalid process parts much earlier than it is possible currently. Moreover, composition does not generate conflicts, thereby improving the effectiveness of the overall approach.

The remainder of the paper is organized as follows. First we describe the underlying formalism and existing work on composition and process validation in Section 2. On this basis we detail how our solution improves these techniques in Section 3. Section 4 discusses related work before Section 5 concludes.

2 Foundations

In order to discuss the concepts in this paper thoroughly, we now introduce the underlying formalism verbally. Subsequently, we discuss shortly today's discovery, composition, and semantic process validation techniques. A running example is used for illustration purposes, and the shortcomings of today's techniques are discussed at the end of the section.

2.1 Semantics for Business Process Models

In the following, the basic execution semantics of the control flow aspect of a business process model are defined using common token-passing mechanisms, as in Petri Nets. The definitions used here are basically equivalent to the ones in [20], which extends [19] with semantic annotations and their meaning. For a more formal presentation please refer to [20].

A process model is seen as a graph with nodes of various types – a single start and end node, task nodes, XOR split/join nodes, and parallel split/join nodes – and directed edges (expressing sequentiality in execution). The number of incoming (outgoing) edges are restricted as follows: start node 0 (1), end node 1 (0), task node 1 (1), split node 1 (>1), and join node >1 (1). The location of all tokens, referred to as a *marking*, manifests the state of a process execution. An execution of the process starts with a token on the outgoing edge of the start node and no other tokens in the process, and ends with one token on the incoming edge of the end node and no tokens elsewhere (cf. *soundness*, e.g., [23]). Task nodes are executed when a token on the incoming link is consumed and a token on the outgoing link is produced. The execution of a XOR (Parallel) split node consumes the token on its incoming edge and produces a token on one (all) of its outgoing edges, whereas a XOR (Parallel) join node consumes a token on one (all) of its incoming edges and produces a token on its outgoing edge.

As for the semantic annotations, we assume there is a background ontology O out of which two parts are used: the vocabulary as a set of predicates P, and a logical theory T as a collection of formulae based on literals over the predicates. Further, there is a set of process variables, over which logical statements can be made, again as literals over the predicates. The logical theory is basically interpreted as a rule base, stating e.g., that all monkeys are animals, or that no man may be married to two women. These rules can then be applied to the concrete process variables, e.g., a particular monkey or a particular man. Further, the task nodes can be annotated using *preconditions* (pre) and *postconditions* (post, also referred to as *effects*), which are also formulae over literals using the process variables. A task can only be orderly executed if its precondition is met, then changing the state of the world according to its postcondition. The postcondition states the explicit effects, and together with the current state and the theory we may derive implicit effects, e.g.: if one carries away a table (with the explicit effect being that the table has been moved), and the theory tells us that everything which is standing on some other thing moves with this other thing, and the local state includes a bottle standing on the table, then the implicit effect is that the bottle is moved too. For any set of literals L we refer to $\overline{L}$ as the union of L and this implications of L from the theory (e.g. $\overline{\mathrm{post_i}}$ is the union of the explicit and implicit effects of T_i).

The question to which degree an explicit effect triggers implicit effects, i.e., further changes the state of the world, yields the well-understood frame and ramification problems. We deal with it by following Winslett's possible models approach [22], resulting in a notion of *minimal changes*, which can be described as a kind of local stability: if there is no indication that something changed, then we assume it did not change. Further, we assume that all changes (unless explicitly inconsistent with the theory) trigger all applicable implications from the theory directly. For more details please refer to [20].

Example 1. *We motivate our solution based on the example process in Fig. 1, represented in BPMN [11]. In this process, every Task T_i is annotated with precondition* $\mathrm{pre_i}$ *and postcondition* $\mathrm{post_i}$. *As outlined above, these annotations allow for an automatic*[3] *transformation into an executable orchestration of Semantic Web Services and for validating the consistency of the process model.*

Let us assume that the ontology contains the literals haveCar, poor, rich, paysBills, billsPaid, haveProgram, *and that the theory says that being rich and being poor are mutually exclusive as well as that iff you are rich you usually pay your bills*[4]*:*

- $\forall x : \neg\mathrm{rich(x)} \vee \neg\mathrm{poor(x)}$,
- $\forall x : \mathrm{rich(x)} \Rightarrow \mathrm{paysBills(x)}$, *and*
- $\forall x : \mathrm{poor(x)} \Rightarrow \neg\mathrm{paysBills(x)}$.

[3] In the ideal case, full automation is possible. In the general case the transformation will be semi-automatic.

[4] We refer to process variables as concrete entities, e.g., me, whereas the variables in the services are not bound yet, e.g., x.

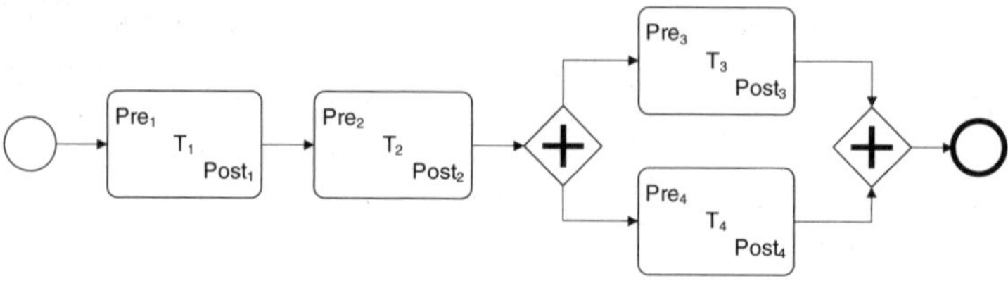

Fig. 1. Example BPMN process

Say, task T_2 is the task of selling your car, and thus annotated with the precondition $\text{pre}_2 := [\text{haveCar}(\text{me}) \wedge \text{poor}(\text{me})]$ *and the postcondition* $\text{post}_2 := [\neg\text{haveCar}(\text{me}) \wedge \text{rich}(\text{me})]$, *where* me *is a process variable. Then, the theory allows us to derive the following implicit effects:* $\neg\text{poor}(\text{me}) \wedge \text{paysBills}(\text{me})$, *and*
$\overline{\text{post}_2} = [\neg\text{haveCar}(\text{me}) \wedge \text{rich}(\text{me}) \wedge \neg\text{poor}(\text{me}) \wedge \text{paysBills}(\text{me})]$.

Based on the execution semantics, we are now able to define the discovery of Semantic Web Services, Service Composition, and Semantic Process Validation more formally.

2.2 Service Discovery

By service discovery we mean the search for a service which matches a requirement best. Semantic Web Service discovery relies on the following input: (1) an ontology O with T and P as above, (2) a set Web services S with their preconditions and postconditions using the predicates in P, (3) a search request Q for a set of Web services which is described through its preconditions and postconditions just as the Web services. Given this input, semantic service discovery returns a set of services R that match the search request Q given ontology O. Three levels of matching are possible [12]: *exact* and *subsumption* matches may suffice to implement a task by a single Web Service; *partial* matches do not completely satisfy the request, however, a combination of them might do.

Example 2. *Consider task T_4 in Fig. 1 which is supposed to implement the "PayBill" action. Task T_4 is annotated with precondition* $\text{pre}_4 := [\text{paysBill}(\text{me})]$ *and postcondition* $\text{post}_4 := [\text{billPaid}(\text{me}) \wedge \text{poor}(\text{me})]$. *Assume that our service repository contains the Web service* CreditCardPayment *which is annotated with* $\text{pre}_{\text{PayBill}} := \text{paysBill}(\text{x})$ *and* $\text{post}_{\text{PayBill}} := \text{billPaid}(\text{x}) \wedge \text{poor}(\text{x})$.

Service discovery with pre_4 *as search request is invoked to find all Web services whose preconditions hold. In this example, the Web service exactly matches the tasks precondition. Next, we check if this service establishes the postcondition of task T_4, i.e. we have to test if the the Web Service's postcondition subsumes the tasks postcondition. In this example, this condition holds, and thus the service* CreditCardPayment *can be selected as the implementation of task T_4.*

2.3 Service Composition

Service composition, in contrast, tries to find a composition of services which jointly satisfy the requirements. In many cases service composition will be performed when service discovery cannot find a single Web service – or it may be that the composition is a better match to the request than any service individually. Several algorithms for service composition exist and can be integrated into our framework, e.g. [4,8].[5] All of them have in common that they find a sequence[6] of Web services $\mathrm{WS_{i_1}}, \mathrm{WS_{i_2}}, \ldots \mathrm{WS_{i_n}}$ for a task T_i where the following conditions hold: (1) the precondition of each Web service is fulfilled at the point where this service is to be executed, i.e., if s is the state in which $\mathrm{WS_{i_j}}$ is be executed, then $s \models \mathrm{pre_{WS_{i_j}}}$ – in particular, in the context here the precondition of T_i enables the execution of the first service, $\mathrm{pre_i} \models \mathrm{pre_{WS_{i_1}}}$; (2) the postcondition of task T_i is satisfied by the state s after the last Web service, i.e. $s \models \mathrm{post_i}$. Notice that for the special case of $n = 1$ the composition result is a subset of the result of service discovery, R.

Example 3. *In the example process in Fig. 1, task* T_3 *may be annotated with* $\mathrm{pre_3}{:=}[\mathrm{rich(me)} \wedge \neg \mathrm{haveProgram(me)}]$ *and* $\mathrm{post_3} := [\mathrm{haveProgram}]$. *Say that there are, amongst others, two services available:* buyComputer *and* writeProgram *with*

$\mathrm{pre_{buyComputer}} := [\mathrm{rich(x)}]$,
$\mathrm{post_{buyComputer}} := [\neg \mathrm{rich(x)} \wedge \mathrm{haveComputer(x)}]$,
$\mathrm{pre_{writeProgram}} := [\mathrm{haveComputer(x)}]$, *and*
$\mathrm{post_{writeProgram}} := [\mathrm{haveProgram(x)}]$.

Note that the resulting composition of Web services contains the literal $\neg\mathrm{rich(x)}$ *as part of its state. In fact, this is a non-obvious inconsistency which semantic process validation, as described below, would detect. When service composition is performed in isolation for every task, inconsistencies like this one can be the result.*

2.4 Process Validation

As mentioned above, a process validation step is needed to detect inconsistencies that may result from performing service composition in isolation for every task. Service composition locally does not lead to any inconsistencies at execution time. Hence, process validation only needs to consider tasks that potentially are executed in parallel because the effects of a task T_1 executed in parallel to another task T_2 may violate T_2's precondition. The basic steps for detecting these inconsistencies are the following [20]:

Computing the parallelity relation. For every task, determine the set of tasks that are potentially executed in parallel. Therefore, a matrix is

[5] Note that there is another notion of service composition, namely the composition of complex behavioral interfaces of a small number of known services – see, e.g., [6,13].

[6] In general there may be more complex solutions than pure sequences – for the sake of brevity we omit further details.

computed that states which pairs of tasks T_i, T_j may be executed in parallel ($T_i \parallel T_j$).

Detection of conflicting parallel tasks. Two parallel tasks $T_i \parallel T_j$ are in *precondition conflict* if there is a literal l such that $l, \neg l \in \mathrm{pre_i} \cup \overline{\mathrm{post_j}}$, or in *effect conflict* if there is a literal l' such that $l', \neg l' \in \overline{\mathrm{post_i}} \cup \overline{\mathrm{post_j}}$.

Determining if the preconditions will always hold. By propagation over the process model it is possible to compute the intersection of all logical states which may exist while an edge e is activated (i.e., a token resides on the edge). This intersection, which we refer to as $I^*(e)$ captures the logical literals which are *always* true when e is activated. Say, T_i is the task node whose incoming edge is e. Then we can check if the precondition of T_i will always hold when T_i may be executed, formally: $I^*(e) \models \mathrm{pre_i}$. If not, we know that the process can be executed such that the precondition of T_i is violated. We refer to this property as *executability*.[7]

Example 4. *Resuming with the example process in Fig. 1 after discovery and service composition was performed as described above, process validation produces the following results: First, the parallelity checker derives that Web service* PayBill *is executed in parallel with Web services* buyComputer *and* writeProgram, *and thus these two Web services can potentially be in conflict with Task* PayBill. *Next, we compute the states before and after executing these Web services. Among these states, state construction will derive the state after executing Web service* buyComputer *to be* $\mathrm{post_{buyComputer}}' :=$ $[\neg\mathrm{rich(me)} \wedge \mathrm{haveComputer(me)} \wedge \mathrm{poor(me)} \wedge \neg\mathrm{paysBill(me)}]$. *Clearly, this conflicts with the precondition of Web service* PayBill. *Notice that this inconsistency is not detected during service composition because it is performed for every task in isolation.*

2.5 Shortcomings of State-of-the-Art Solutions

The sequence of steps described in this section implies a number of limitations and problems that we address with our solution. First, service discovery and composition can only exploit the annotations explicitly attached to the tasks in the process. Consider the example of a task T_i with $\mathrm{pre_i} := \mathrm{poor(me)}$ and $\mathrm{post_i} := \mathrm{haveProgram(me)}$. Then neither service discovery nor composition would find any applicable services. If, however, the previous step in the process was buyComputer, and the computer has not been sold already, then there is an additional (hidden) precondition haveComputer(me). Taking this precondition into account makes the problem solvable through discovery and composition. Second, as shown in Section 2.3 service composition is performed for a single task in the process in isolation because using process level composition would be close to automatic programming, and thus computationally prohibitive. The resulting orchestration of services is locally consistent and correct. However, inconsistencies may be caused by a lack of information about dependencies to other

[7] Note that this this is a design time property of a process, which indicates if or if not instances of this process may not execute due to preconditions that are not fulfilled.

tasks. Therefore, a separate validation step is required to detect inconsistencies between the orchestration computed for different tasks in isolation. Third, inconsistencies induced during service composition per task or due to inconsistent annotations are detected very late in the whole procedure, wasting computing resources and slowing down the modeling process.

3 Solution Approach

In this section we present a detailed solution to the shortcomings identified in the previous section. The rough outline of the approach is the following: (1) before starting discovery or composition, all implicit information is derived from the process context; (2) based on this information we can potentially extend the preconditions of tasks that are to be implemented; (3) also due to the context information, the annotations of tasks are enriched with constraint-sets which are used to avoid the construction of compositions that are inconsistent with the rest of the process model. Given the extended preconditions and postconditions resulting from (1), service discovery and composition are able to detect more candidate services which increases the chance that any solution is found.

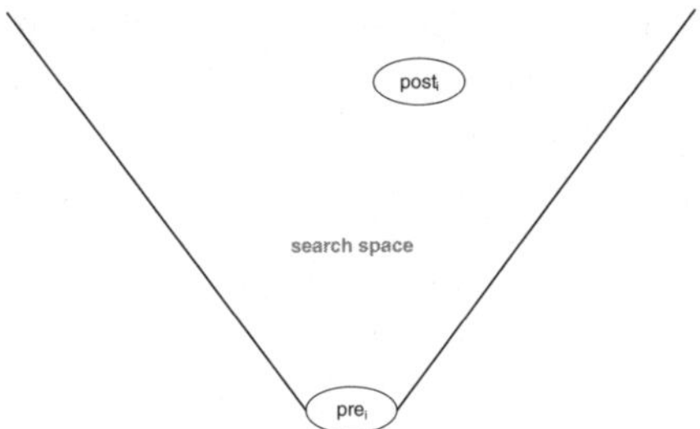

Fig. 2. Original search space

In the remainder of this section we focus on the effects of these extensions on the search space in service composition. This trivially relates to discovery by considering single-step compositions only. In order to better explain the implications, we visualize the composition search space. Fig. 2 shows an exemplary, unchanged search space, where the search starts from pre_i and tries to reach post_i.[8] The funnel above pre_i then indicates the boundaries of the search space. This visualization of the search space serves as an intuitive and informal way to explain the idea of our approach. For a more formal treatment of search space implications, we refer to, e.g. [4,20]. Possible solutions can be depicted as paths from pre_i to post_i, as depicted in Figure 3.

A solution path is only valid if it does not cross the boundaries of the search space. Note that Path 1 in Fig. 3 leads through a region which is out of boundaries once the constraint-set is taken into account. In other words: Path 2 is valid

[8] While the illustrations are related to forward search, the techniques can be applied analogously to backward search. For these basic planning techniques see [15].

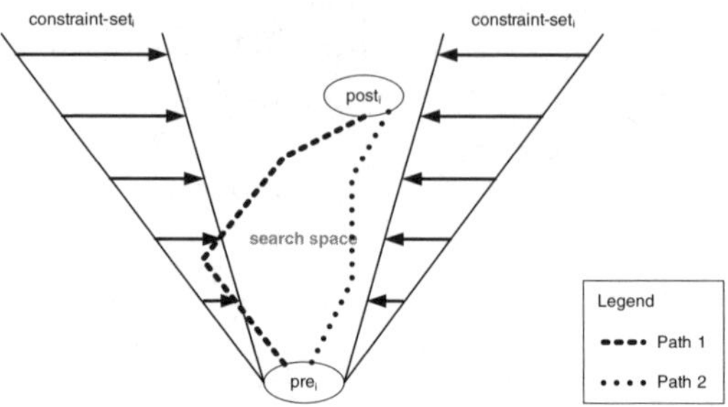

Fig. 3. Alternative solution paths, one of which violates the constraint-set

with and without considering the constraint-set, whereas Path 1 is only a valid solution when the constraint-set is neglected.

3.1 Modification of the Search Space

In Section 2, we explained how current solutions compute a possible solution given the precondition and a postcondition associated with a task (cf. Fig. 2). Below, we show how the expansion of the precondition and the inclusion of constraint-sets affects the shape of the search space.

First, we propose to expand the precondition of every task. For this purpose we merge its precondition with the logical facts we know must always hold when this task may be executed (given through I^* of the incoming edge, see Section 2.4). Basically, those are the literals which were established at one point in the process before the current task, i.e., that were true from the start on, or that were made true in the postcondition of another task, and which cannot be made false before the task at hand is executed. The computation of this set of literals can be done in polynomial time for our restricted class of processes (in the absence of loops), e.g., by an efficient propagation technique such as the I-propagation method described in [20].[9]

As shown in Fig. 4, our algorithm expands the precondition pre_i. In general, the extended precondition allows us to discover additional applicable Web services, because discovered Web Services can rely on more known preconditions. This increases the choices to orchestrate Web Services, and thus it becomes more likely to find any valid orchestration of Web Services. In Fig. 2, imagine that the goal, i.e. the postcondition post_i, was not completely inside the search space, but that in Fig. 4 it was. This would mean that a previously unsatisfiable goal becomes satisfiable by using the precondition expansion.

The constraint-set, our second extension to current solutions, may have an opposite effect. Unlike preconditions and postconditions, which are conjunctive formulae, a constraint-set is a set of literals interpreted as a disjunctive formula

[9] This algorithm is implemented and runs in well under a second for common examples. Given its polynomial time worst-case behavior, scalability is unlikely to be a problem.

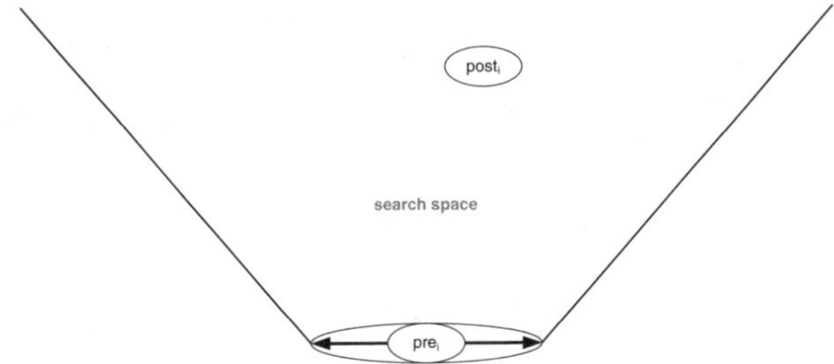

Fig. 4. Search space with expanded precondition

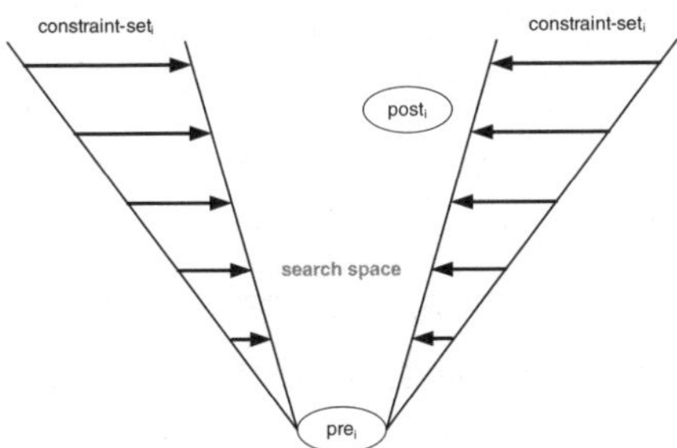

Fig. 5. Constrained search space

which expresses constraints on the states that may be reached during the execution of a task: if one of the literals from the constraint-set appears, the constraint is violated. The constraint-set is used to preemptively avoid parallelism-related conflicts. Thus, it is computed as the negated union of all preconditions and postconditions of the tasks that may be executed in parallel to the chosen task node n_i: constraint-set$_i := \bigcup_{n_j \| n_i} \{\neg l \mid l \in \mathrm{pre_j} \cup \mathrm{post_j}\}$.

Besides constraints explicitly modeled by the user (e.g. only services provided by preferred service providers may be considered), we can apply the M-propagation algorithm presented in [20] to compute the set of parallel nodes for each task node. Given that, the computation of a constraint-set$_i$ is straightforward. As above, this can be done in polynomial time. Note that it is a deliberate choice to use a restricted logical formalism. While richer formalisms offer more expressivity, the here proposed extensions for modeler support are to be used during process modeling. Since long waiting times are undesirable, we propose a restricted formalism here.

As shown in Fig. 5, constraint-sets restrict the search space considered during service composition. As an important positive implication service composition avoids generating solutions that will conflict with tasks executed in parallel. This can be achieved by filtering out services whose preconditions and postconditions would cause a conflict with the constraint-set in a filtering step right before the

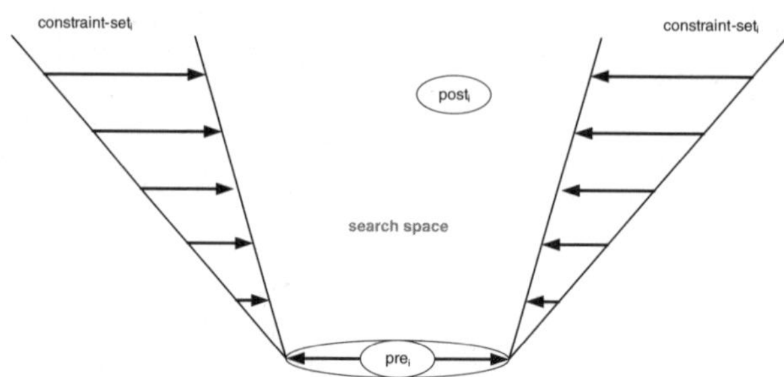

Fig. 6. Constrained search space with expanded precondition

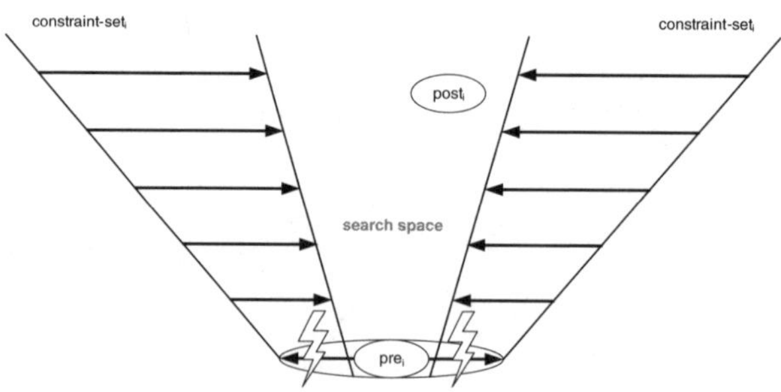

Fig. 7. Constrained search space with expanded precondition, and conflicts between the precondition and the constraint-sets

actual composition. In effect, invalid orchestrations of Web Services will not even be generated and thus need not be detected (and fixed) later – e.g., as in Fig. 3.

By including both extensions (Fig. 6) we gain the best of both: while considering further relevant services we restrict the search space to valid compositions.

Fig. 7 depicts another interesting case: here, the expanded precondition is in conflict with the constraint-set. Apparently, the respective goal is not satisfiable and should not be considered. Note that in this case the conflict is between the expanded precondition and the narrowed boundaries of the constraint-set. However, the conflict can also be present between the original boundaries and the original precondition, or any combination in between.

Example 5. *Reconsidering the example from Section 2, we now indicate how the information gathering approach described in this section improves business process modeling. First, based on expanded preconditions more relevant services can be discovered, specifically when a service relies on postconditions established by predecessors of a task which cannot be derived from the precondition of a task alone. This improves upon the example mentioned in Section 2.5 for current solutions. Second, service composition does not compute an invalid orchestration here when it takes into account the constraint-sets. Thus, in Example 3 service composition in our method would not even generate a solution in which the implementation of task* T_3 *is in conflict with task* T_4*.*

3.2 Configuration Options

An interesting aspect of our approach to semantic business process modeling is its adaptability to specific user needs along three dimensions: (1) With our method the user can, to a degree, trade completeness of the solution for efficiency. (2) The quality and granularity of error messages can be configured to the needs of the modeler. (3) It is possible to examine more suitable compositions beyond the initial solution found by service composition.

Completeness vs. Efficiency. First, the user trade-off on completeness of composition vs. efficiency: one can configure the modeling tool such that preconditions of tasks in a process are not expanded, leading to fewer Web services being considered during composition and composition is performed faster (it is exponential in the number of candidate services [4]). However, service composition may not find a valid composition of tasks that implements the modeled process even if this is possible with the available Web services (compare Fig. 2 and Fig. 4).

Error Reporting. Second, consider the problem that service composition fails because a task in the process model cannot be composed without violating its constraint-set. Then it may be useful to actually use the composition regardless of the violation, and change other parts of the process model to resolve the inconsistency. This becomes possible, e.g., by expanding the precondition but ignoring the constraint-set of a task. An explicit semantic process validation step, as it is used by state-of-the-art approaches, then highlights the violation. Thereby, error messages presented to the user can convey more information that help him to adjust the process.

Quality of the Composed Process. Finally, assume that a valid composition is already found by not expanding preconditions. One might still be interested in a more suitable service composition, if any exists that is to implement if preconditions are expanded.

These three dimensions for configuring the process underline the power of being able to adapt the process to the specific needs of the process modeler, and hence are another contribution of this paper.

3.3 Applying Search Space Modifications During Modeling

The modification of the search space outlined in this section is embedded into the overall translation procedure that we describe below. Our integrating solution extends the one presented in [21] while sharing its basic architecture.

Figure 8 summarizes the steps we follow for modeling a process. The user triggers service composition for an arbitrary task in the modeled business process. This can be done explicitly by a function invocation or implicitly, e.g. after the modeler has finished to define the task. As a result, a request is sent to to derive all relevant information, which includes for each task, one after the other finding all states that can be executed in parallel, expanding the precondition, deriving the constraint-set after which service discovery and service composition is performed. This is repeated for each of the tasks in an order that reflects the

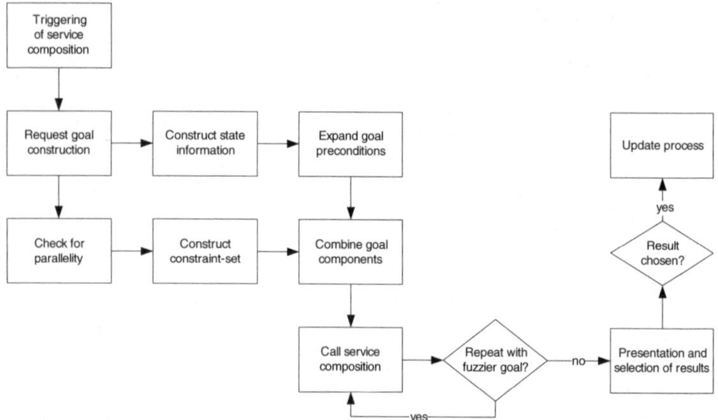

Fig. 8. Extended composition procedure

control flow over the tasks. Depending on whether or not the constraint-set is taken into account, the process needs to undergo semantic validation as a final step again. When no composition of the services was found, or when the modeler is not satisfied with the solution, the conditions attached to the tasks of the process model can be relaxed as discussed in Section 3.2.

4 Related Work

In this paper we present a solution that integrates service composition with process validation to improve the efficiency of business process modeling. While to the best of our knowledge no integrated approach has been proposed yet, isolated solutions for service discovery, service composition, or process validation may be reused as building blocks in our integrated solution. The general approach and architecture that underlies our solution extends [21], which delivers implementation and configuration support in Semantic Business Process Management.

Our approach integrates well with the existing discovery mechanisms for Semantic Web Services, e.g. [7,12] which are based on subsumption reasoning. In order to make our solution more flexible in the presence of erroneous business process models, we may extend service discovery to approximate reasoning. The feedback provided by the discovery mechanism outlined in [17] may provide guidance in case of empty results during service discovery, which can result from inconsistent use of the background ontology or distributed ontologies.

The methods for service composition [4,8,10] and validation [20] referred in this paper are based on the possible models approach of AI planning [22]. In principle, other algorithms or formalisms for discovery [1], composition [2,10,14,16], or validation [5,9] may be used or adapted to this context.

However, note that [9] uses a fundamentally different way to check the semantic consistency of process models: their work yields a checking method for pairwise constraints over task co-occurrences (e.g., task A must always be preceded by task B) in block-based processes. In contrast, our underlying semantic process validation checks if declaratively annotated preconditions and postconditions are consistent with a flow-based process model and a background ontology. This is achieved by our I-propagation, which propagates summaries of the logical states that may be encountered. And exactly this logical state information then is used in the work presented in this paper for the precondition expansion – which disallows relying on [9] for the purposes here.

With respect to [5], it is worthwhile noting that their approach goes into a similar direction of propagating effects. Besides some formal flaws and concerns about the computational properties of their work, further differences lie in the focus on the interoperation of processes in [5] and the absence of a thorough consideration of background ontologies.

5 Conclusion

In this paper we presented a method to support the modeler of a semantic execution-level business process. For this purpose, the available semantic information in the process is collected and made available to service discovery and composition, leading to semantically consistent compositions. This is achieved by integrating previous work on service composition and process validation. In more detail, we use the process context to (i) extend the precondition which is certain to hold at a given point in the process, and (ii) derive a constraint-set, i.e., constraints on the intermediate states that may be reached inside a process activity (e.g., by service composition).

In terms of runtime, these extensions have contrary effects: on the one hand, the search space considered during service composition is reduced through the constraint-sets, as they are used to prune away solutions which would lead to inconsistent states in the resulting orchestration anyhow; on the other hand, the extended preconditions may lead to a larger search space, if the composition is performed in the manner of a forward search (i.e., starting at the precondition, searching towards the postcondition); if, however, composition is done in a backward search manner (i.e., in the opposite direction), this downside can be avoided.

While in practice the transformation of business processes into an executable process is a largely manual task, our solution is a clear step towards better automation. Thus, based on the presented solution the realization of changes to business processes may experience a significant speed-up. In future work, we plan to verify our claims experimentally by integrating and extending our previous work [4,20] as described here.

References

1. Akkiraju, R., Srivastava, B., Anca-Andreea, I., Goodwin, R., Syeda-Mahmood, T.: Semaplan: Combining planning with semantic matching to achieve web service composition. In: 4th International Conference on Web Services (ICWS 2006) (2006)
2. Constantinescu, I., Faltings, B., Binder, W.: Typed Based Service Composition. In: Proc. WWW 2004 (2004)
3. Hepp, M., Leymann, F., Domingue, J., Wahler, A., Fensel, D.: Semantic business process management: A vision towards using semantic web services for business process management. In: ICEBE, pp. 535–540 (2005)

4. Hoffmann, J., Scicluna, J., Kaczmarek, T., Weber, I.: Polynomial-Time Reasoning for Semantic Web Service Composition. In: Intl. Workshop Web Service Composition and Adaptation (WSCA) at ICWS (2007)
5. Koliadis, G., Ghose, A.: Verifying semantic business process models in interoperation. In: Intl. Conf. Services Computing (SCC 2007) (2007)
6. Lemcke, J., Friesen, A.: Composing web-service-like abstract state machines (ASMs). In: Intl. Conf. Services Computing - Workshops (SCW 2007) (2007)
7. Li, L., Horrocks, I.: A software framework for matchmaking based on semantic web technology. In: 12th World Wide Web Conference (WWW 2003) (2003)
8. Lutz, C., Sattler, U.: A proposal for describing services with DLs. In: International Workshop on Description Logics 2002 (DL 2002) (2002)
9. Ly, L.T., Rinderle, S., Dadam, P.: Semantic correctness in adaptive process management systems. In: Dustdar, S., Fiadeiro, J.L., Sheth, A.P. (eds.) BPM 2006. LNCS, vol. 4102, Springer, Heidelberg (2006)
10. Meyer, H., Weske, M.: Automated service composition using heuristic search. In: Dustdar, S., Fiadeiro, J.L., Sheth, A.P. (eds.) BPM 2006. LNCS, vol. 4102, pp. 81–96. Springer, Heidelberg (2006)
11. OMG. Business Process Modeling Notation – BPMN 1.0. Final Adopted Specification, February 6, 2006 (2006), `http://www.bpmn.org/`
12. Paolucci, M., Kawamura, T., Payne, T., Sycara, K.: Semantic matching of web services capabilities. In: Horrocks, I., Hendler, J. (eds.) ISWC 2002. LNCS, vol. 2342, Springer, Heidelberg (2002)
13. Pistore, M., Marconi, A., Bertoli, P., Traverso, P.: Automated composition of web services by planning at the knowledge level. In: IJCAI (2005)
14. Rao, J., Su, X.: A Survey of Automated Web Service Composition Methods. In: Cardoso, J., Sheth, A.P. (eds.) SWSWPC 2004. LNCS, vol. 3387, pp. 43–54. Springer, Heidelberg (2005)
15. Russell, S., Norvig, P.: Artificial Intelligence: A Modern Approach. Prentice-Hall, Englewood Cliffs (1995)
16. Sirin, E., Parsia, B.: Planning for semantic web services. In: McIlraith, S.A., Plexousakis, D., van Harmelen, F. (eds.) ISWC 2004. LNCS, vol. 3298, Springer, Heidelberg (2004)
17. Stuckenschmidt, H.: Partial matchmaking using approximate subsumption. In: AAAI, pp. 1459–1464 (2007)
18. van der Aalst, W.M.P., ter Hofstede, A.H.M., Weske, M. (eds.): Business process management: A survey. Business Process Management (BPM). LNCS, vol. 2678. Springer, Heidelberg (2003)
19. Vanhatalo, J., Völzer, H., Leymann, F.: Faster and More Focused Control-Flow Analysis for Business Process Models though SESE Decomposition. In: Krämer, B.J., Lin, K.-J., Narasimhan, P. (eds.) ICSOC 2007. LNCS, vol. 4749, Springer, Heidelberg (2007)
20. Weber, I., Hoffmann, J.: Semantic business process validation. Technical report, University of Innsbruck (2008), `http://www.imweber.de/texte/tr-sbpv.pdf`
21. Weber, I., Markovic, I., Drumm, C.: A Conceptual Framework for Composition in Business Process Management. In: Abramowicz, W. (ed.) BIS 2007. LNCS, vol. 4439, Springer, Heidelberg (2007)
22. Winslett, M.: Reasoning about actions using a possible models approach. In: Proc. AAAI 1988 (1988)
23. Wynn, M., Verbeek, H., van der Aalst, W., ter Hofstede, A., Edmond, D.: Business process verification - finally a reality! Business Process Management Journal (2007)

Challenges in Collaborative Modeling: A Literature Review

Michiel Renger[1], Gwendolyn L. Kolfschoten[1], and Gert-Jan de Vreede[1,2]

[1] Department of Systems Engineering, Faculty of Technology Policy and Management, Delft University of Technology, The Netherlands
d.r.m.renger@tudelft.nl, g.l.kolfschoten@tudelft.nl
[2] Institute for Collaboration Science, University of Nebraska at Omaha, USA
gdevreede@mail.unomaha.edu

Abstract. Modeling is a key activity in conceptual design and system design. Users as well as stakeholders, experts and entrepreneurs need to be able to create shared understanding about a system representation. In this paper we conducted a literature review to provide an overview of studies in which collaborative modeling efforts have been conducted to give first insights in the challenges of collaborative modeling, specifically with respect to group composition, collaboration & participation methods, modeling methods and quality in collaborative modeling. We found a critical challenge in dealing with the lack of modeling skills, such as having a modeler to support the group, or create the model for the group versus training to empower participants to actively participate in the modeling effort, and another critical challenge in resolving conflicting (parts of) models and integration of submodels or models from different perspectives. The overview of challenges presented in this paper will inspire the design of methods and support systems that will ultimately advance the efficiency and effectiveness of collaborative modeling tasks.

Keywords: Collaborative modeling, system analysis and design, groups, participation, modeling methods.

1 Introduction

Modeling is a key activity in conceptual design and system design. There is broad agreement that it is important to involve various experts, stakeholders and users in a development cycle [1-3]. While these parties are often interviewed or in other ways heard, they often lack the skills to actively participate in the modeling effort. If users are not involved in systems analysis tasks, their problems, solutions, and ideas are difficult to communicate to the analyst. This often results in poor requirements definition, which is the leading cause for failed IT projects [1].

Further, analysts and entrepreneurs might have mental models, visions of a solution or system design, but might lack the adequate means of articulating these in terms familiar to all stakeholders involved [4]. While there are means to verbally explain models, such as metaphors, a graphical representation is often more effective. ("A picture tells more than a thousand words"). In order to use graphical representations as a basis for discussion, it would be useful if all the stakeholders can be actively engaged in the construction and modification of such models.

J.L.G. Dietz et al. (Eds.): CIAO! 2008 and EOMAS 2008, LNBIP 10, pp. 61–77, 2008.

With increasing complexity of systems and organizations, creating shared understanding and joint representations of those systems becomes increasingly important. Analytical skills become more wanted and more important to function in these complex contexts. However, creating one's own system representation is in many ways different from creating a joint system representation. With the increasing need for collaboration among experts and knowledge workers [5], collaborative modeling becomes increasingly important.

Collaborative modeling has been a research topic since the late 70's [6]. In order to support collaborative modeling to create shared understanding and joint visions for design and solution finding it is important to gain insight in best practices and key challenges in collaborative modeling. Most articles on collaborative modeling describe case studies and practical experiences with collaborative modeling. While meta-analysis has been performed to gain insight in metrics and effects in collaborative modeling [7] and an overview of methods and role deviations has been described [8, 9], to our knowledge there is no overview of challenges and best practices in collaborative modeling. Such overview would help us to find opportunities for research and for the design of new supporting tools and methods to empower participants and facilitators for effective and efficient collaborative modeling.

In this paper we provide an overview on collaborative modeling studies that brings together the experiences and findings from literature to identify the main challenges and lessons learned in the field. This could inspire research on new and innovative collaborative modeling support systems and methods. Furthermore such overview of challenges will be a valuable resource for practitioners in collaborative modeling. The paper first defines collaborative modeling. Next we describe the different approaches in collaborative modeling. Third, we discuss the research method for the literature review, followed by the results in which we describe critical challenges and solutions for successful collaborative modeling. We end with conclusions and suggestions for further research.

2 Background

2.1 Collaborative Modeling Defined

For the purpose of the research presented in this paper, we define collaborative modeling as:

The joint creation of a shared graphical representation of a system.

In this definition we focus on <u>graphical</u> modeling of systems as opposed to physical modeling of objects or artifacts such as in architecture and industrial design. Graphical models are usually created in a conceptual design phase either for analysis or design. They are used to communicate a representation of a system in order to understand or change it. Conceptual models are used in a early phase of analysis and design and therefore are initially associated with a sketching activity. However, when they are used as a basis for design or structural analysis they need to meet various requirements with respect to precision and rigor. They also may need to be translated to computer models in order to calculate effects of the model. For this purpose, modeling languages have been developed to capture conceptual models as computer models to enable easy manipulation and automatic syntactic verifications. Using computer models also makes it easier to make changes in a model, especially when changes in one component result in changes to other components and relations between components.

Further, we focus on joint creation to indicate our interest in stakeholder participation in the modeling effort as opposed to modeling by external professionals or analysts only. Joint creation requires the exchange of perspectives among the participants. The model is a way to elicit, highlight, and communicate different perspectives and assumptions among group members.

In order to create a shared representation as opposed to an individual representation, a shared understanding of the elements and relations in the model needs to be created. Shared understanding can be defined as "the overlap of understanding and concepts among group members" [10, p. 36]. We build on this definition for the collaborative modeling domain where we define shared understanding as *the extent to which specific knowledge among group members of concepts representing system elements and their relations overlaps.* In order to create overlap in knowledge, participants need not only share information about model elements and relations. They also need to create shared meaning with respect to these elements and their relations. Creating shared meaning is often studied from a 'sensemaking' perspective. Sensemaking is described by Weick as involving "the ongoing retrospective development of plausible images that rationalize what people are doing" [11, p. 409]. Sensemaking usually requires some development of shared meaning of concepts, labels, and terms. It also includes the development of a common understanding of context and the perspective of different stakeholders with respect to the model.

2.2 Approaches in Collaborative Modeling

Within the field of system dynamics, modelers started to involve client groups in the modeling process since the late 70's [6]. Since that time, various other modeling approaches have adopted the notion of collaborative modeling and found methods to involve stakeholders in their own modeling efforts, see e.g. [8, 12, 13]. As a result various research groups performed field studies, gained experience, and eventually developed sophisticated methods for modeling efforts that have high levels of stakeholder participation. However, different modeling languages are associated with different methods for analysis and design. To accommodate different stakeholder groups, new methods had to be developed leading to different approaches and eventually different patterns in collaborative modeling. Here we describe the most important approaches in collaborative modeling.

2.2.1 Problem Structuring Methods

Problem Structuring Methods refer to a broad variety of methods and tools that have developed mainly in the UK to cope with complexity, uncertainty and conflict [14]. The most well known of these methods are Soft Systems Methodology, Strategic Choice, and Strategic Options Development and Analysis (SODA), of which the last has further developed into Jointly Understanding Reflecting and Negotiating Strategy Making (JoURNeY Making) [15]. These methods have the following similarities in approach: use of a model as a transitional object, increasing the overall productivity of group processes, attention to the facilitation of effective group processes, and significance of facilitation skills in enabling effective model building [16]. Especially the first of these is characteristic for Problem Structuring Methods: models are seen as instrumental to strategic decision making and problem solving in complex settings, so the approach typically focuses on the overall decision making process which often includes simulation for scenario explorations. In Problem Structuring Methods, a group's shared understanding is created by switching between the views of individual participants and the entire group, and focus on the differences in view to resolve these [17, 18].

2.2.2 Group Model Building

Group Model Building is considered a special case of Problem Structuring Methods for hard modeling, and has been developed by researchers of the University at Albany in New York, and the University of Nijmegen in the Netherlands. According to Andersen et al, Group Model Building refers to "a bundle of techniques used to construct system dynamics models working directly with client groups on key strategic decisions [19, p. 1]". Group Model Building always has a system dynamics approach, and usually extends the conceptual model to simulation models to explore diverse strategic options. The flexible outlines of the method, the so-called scripts, are presented in [8].

The more applied and specific approach called Mediated Modeling is largely based on these scripts and is developed mainly for complex problems of ecological nature. Shared Vision Modeling provides a similar approach to handle water resource management problems [20].

2.2.3 Enterprise Analysis

Originally, collaborative modeling research at the University of Arizona has a stronger focus on the development of software tools as well as facilitation techniques for the support of collaborative modeling efforts [21, 22]. This approach concentrates more on collaboratively built models as a goal in itself rather than as a transitional object. The models to be built are often of the IDEF0 type, and many techniques are especially meant to deal with IDEF0-standard models [23, 24].

Apart from these above described approaches, the term Participatory Modeling is loosely used for collaborative modeling across different approaches. Companion Modeling (ComMod) uses multi agent systems or role playing games to elicit information needed to construct the model [25]. The used methods are designed to build different types of models in different contexts with different purposes. In this article we focus on the similarities between modeling methods and challenges in the collaborative modeling effort.

3 Method

In this paper we focused on the challenges that groups encounter when they engage in a collaborative modeling effort. As a research method we used a literature analysis. Because the data collected in collaborative modeling is often of a qualitative nature, a quantitative meta-analysis was not feasible. Moreover, a structural survey among key journals would have been highly inefficient as collaborative modeling research can be found in journals concerning modeling and collaboration as well as in domain specific journals such as "Water Resources Update". While the resulting set of articles might not be a complete set, the interdisciplinary nature suggests that a more qualitative approach is most efficient and effective to create an overview of existing literature on collaborative modeling. In total we found 46 papers.

We analyzed papers that studied collaborative modeling in which the deliverable is a graphical model of a system, as discussed in the definition. We searched for articles in which the modeling method was the central topic. We did not discriminate among research methods or approaches [21, 26, 27]. To identify articles we searched in various research databases such as Google Scholar, Elsevier's Scopus, IEEE Explorer, the ACM Portal and Science Direct on: collaborative modeling/ modelling, participatory modeling, group model building, shared vision modeling, and mediated modeling. Further, we searched for articles on collaborative modeling within the context of the related subjects such as: facilitation, G(D)SS, Collaboration Engineering, (Information) Systems and Software Engineering, Business Process Modeling, collaborative design, and collaborative learning. From the papers we found we searched the bibliography for additional references, and we looked at papers that cited the papers we found [28].

For each article we searched for challenges and lessons learned. We captured these in a database and compared the different findings from different perspectives. We based these perspectives on the framework used by Nunamaker et al. to study the effects of electronic meeting systems [29] (see figure 1).

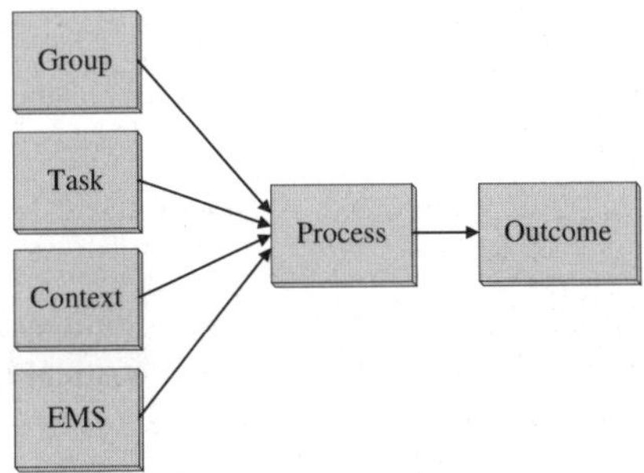

Fig. 1. Framework used by Nunamaker et al. (1991) to study effects of EMS

We adapted this framework as to reflect the most relevant factors of collaborative modeling as found in the literature:

- Group. We focused on the composition of the group and the different roles that can be present during a modeling effort.
- Task and Context. We found papers in a wide variety of (possibly overlapping) domains, including Systems Theory (15), technological support and Software Engineering (14), collaboration and facilitation (11), Business Process Modeling (5) and environment management (3). We also found that the collaborative modeling occurred in various organizational settings such as public (7), military (4), insurance (2), health

care (2), and software engineering (2). Due to this broad scope, we chose to perform our analysis independently of the task and contextual factors like domain and organizational setting in which the modeling method is used.
- EMS. We found several studies in which specific tools were used to support collaborative modeling and in which this technology was the central topic. However, this paper focuses on the modeling effort, not on the technological design of supporting tools. Therefore, we study the technological aspect in terms of its functionalities only, and how they relate to other aspects.
- Process. Since we are interested in the interaction challenges of participants and the role of facilitation support in collaborative modeling efforts, we studied the process from two perspectives: the interactive process and the modeling method.
- Outcome. For collaborative modeling, the model quality is of specific interest. Factors like efficiency and buy-in are considered in relation to other aspects.

Our tailored framework thus consists of four key aspects of collaborative modeling, as shown in figure 2. The results are described and compared in the next section.

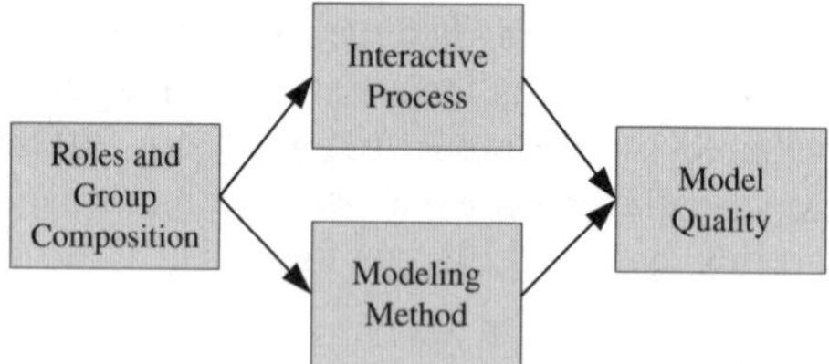

Fig. 2. Adapted framework with four key aspects of collaborative modeling

4 Results

As discussed in the Method section, we study the challenges and lessons learned found in the literature within four topics that represent critical choices in the design of a collaborative modeling activity:

1. The roles and group composition
2. The interactive process; collaboration and participation
3. The modeling method, activities and modeling rules to support the modeling effort
4. The model quality, both from an objective and a subjective perspective.

4.1 Roles and Group Composition

Collaborative modeling requires expertise in two distinct area's; facilitation of the group process and expertise in modeling, and the modeling semantics. Such expertise is generally not available in organizations and is therefore often outsourced. On the other hand, participants can also fulfill different roles in the collaborative modeling process to coordinate tasks and responsibilities in the modeling effort.

4.1.1 Facilitation Roles

Richardson and Andersen have described five essential roles that should be present in a Group Model Building session: the facilitator, the modeler/ reflector, the process coach, the recorder and the gatekeeper [9]. Roles can be allocated to different people in the group, which can effect the workload of those participants and therewith the effectiveness and the efficiency of collaboration support [30]. Furthermore, some roles can be combined, or even (partly) assigned to group members. Having an outside facilitator is considered very useful, especially if technology is used [31, 32]. Vennix et al. note that facilitated groups get less frustrated, have strongly improved group performance, less social-hierarchical domination in discussions and focus on a broader spectrum of approaches to the problem [6].

In traditional modeling methods, the input of stakeholders is processed into a model by the analyst/ modeler. But also in more collaborative settings the role of modeler is mentioned in the literature. However, a modeler/ reflector will not only support the process of collaborative modeling, but will also interfere with the content to help groups to understand the system or process under discussion. There is a discussion among scholars about the effect and ethics involved in content interference by outside facilitators [33]. Also, there is no consensus among scholars about whether the roles of facilitator and modeler should be represented by separate persons. Especially when the task is complex, a large cognitive load is imposed on a person that serves both [9]. Separated roles of modeler and facilitator are found to save time and increase model quality [9, 34, 35]. A facilitator and modeler need to work together seamlessly and be careful not to create conflict between each other. Van den Belt suggests that these roles are therefore inseparately intertwined, and a combined role of facilitator and modeler is equally or even more efficient because it allows a stronger focus on conflict resolving tasks [36]. Moreover, a larger supporting team is more expensive. A third possibility would be to assign the modeler role to the group of participants as a whole, which would increase the participation of group members, but would also pose great challenges for the facilitation. Little explicit research is found with the use of this approach.

The recorder, also known as chauffeur provides the technical support to the group by processing the input of the group directly into a modeling tool and by operating any additional technology such as group support systems [9, 30]. The role of the chauffeur is closely related to the role of modeler, but the chauffeur functions more as a scribe, whereas the modeler also interprets and reflects on the group's input. Both roles can be executed by one person.

The process coach focuses solely on the dynamics of individuals and subgroups and serves as an assistant to the facilitator to decrease cognitive load. The role of the process coach is to detect conflict, uneasiness, dissatisfaction, a lack of motivation and other signs of the group that require action from the facilitator. It is the task of the process coach to not only identify these needs, but also to suggest remedies [9]. Little literature is found that explicitly mentions this role.

Finally, the gatekeeper is the medium between the facilitation and the participation roles. This role is a twofold representative: for the participants he represents the supporting team and vice versa. Usually this is a person within the organization who initiates and carries responsibility for the project. The gatekeeper can help the facilitation team in preparation tasks and in assigning participation roles [9].

4.1.2 Participation Roles

One of the main reasons for failure of business process re-engineering projects is the involvement of the wrong people at the wrong time for the wrong reasons [34]. Within all approaches the importance of selecting the right participants is acknowledged [8, 37]. There are several factors that should be kept in mind when composing the group. Critical choices should be made with respect to the involvement of experts (professionals or people with experience) and the involvement of stakeholders. Sometimes both experts and stakeholders are involved to achieve both model quality as well as support for the resulting representation of the system or process.

With respect to expert involvement a trade-off emerges between quality of the model and shared understanding. First of all, the richness of the expertise in the group should be considered in order to produce a complete model that covers the scope of the system. However, when experts have non-overlapping expertise, it might become more difficult to create shared understanding.

A similar trade-off occurs when inviting stakeholders. When critical stakeholders are not invited, the group can have insufficient decision power, and there can be a lack of support for solutions and decision by non-invited or insufficiently heard stakeholders [34, 38]. Dean, Lee et al. note that the presence of both process designers and process owners is important in process analysis [22]. In other methods user involvement might be critical. On the other hand, more stakeholders can result in more conflict, which will require more consensus building activities. For the same reason it is found more difficult to maintain motivation among participants in a freestanding project with non-professional interests [36].

In general, in a large group participants have less average speaking time in "live" discussions [35]. Further, a large group produces a lot of information, which in its turn puts a strain on the cognitive load of individual participants with the danger of information overload [23, 24].

Although there are means to overcome the challenges of large groups, e.g. by working in parallel by the use of group support systems, group discussions are found to be at the heart of Group Model Building processes [35]. In some cases, different stakeholders or experts are involved at different steps in the modeling process to reduce the burden on the costly time of professionals. However, stakeholders that are involved later can put earlier made decisions back on the agenda, or even reject these [34, 38]. Simultaneously involved participants are found to increase pace and buy-in [34].

4.2 Collaboration and Participation

As mentioned above, the presence of a modeler or chauffeur has a large impact on the group processes. When using a scribe/ modeler role individual group members have less direct access and power to influence the model. The rules for scribe/ chauffeur and modeler are different. In a chauffeured setting changes made to the model will have to be agreed upon by the group before the chauffeur effectuates the change. Another rule, more associated to the modeler role is that the modeler can freely interpret the group discussion, and based on this, change or extend the model [9].

Another effect of the presence of the chauffeur or modeler is that participants have less access and ability to modify the model, and therewith to explain their perception to the group. This lack of access can lead to less interactivity because of more indirect communication among group members via the chauffeur [35]. This can decrease the feeling of data ownership or group contribution to the model. Also, some cases in which a modeler carried out the steps were identified as time-consuming [34]. In some cases a modeler interpreted and made changes to the model in between sessions which resulted in feelings that the model no longer captured the group's original intent [13]. A problem with inability to change the model is that on one hand, participants are asked to 'translate' their perceptions and ideas to the modeling language, but on the other hand they do not get the opportunity to verify this translation, or to express their perception in the common modeling language. This lack of ability to express and the lack of feedback can cause a feeling of not being understood, and not being able to express a vision or perspective.

Experiments with a modeler making changes simultaneously with the group were less successful because the modeler could not keep up with the pace of the group. Another possibility is that the group can directly make changes to the model, while the facilitator or a separate modeler is present to give modeling guidance to the group [13, 22]. In most cases of unchauffeured modeling, group members were given a modeling training in advance of the session, which is acknowledged to be critical [23].

A critical enabler of full group participation is the ability to work in parallel. In all cases were the model was built in parallel, the group divided into subgroups and subgroups were assigned parts of the model corresponding to the subgroups' expertise. Dennis et al. found that parallel built models are built ten times as fast as models built with the plenary group [23]. While in general facilitation with group support systems individual parallel work is common, little literature is found on individual parallel model building. In both cases a key challenge lies in change management and change awareness among parallel working individuals or groups. Andersen and Richardson found that convergent thinking requires the input of the group as a whole [6, 8].

Group processes of collaborative modeling efforts can be strongly affected by social factors that are apparent among group members, e.g. organization-hierarchical factors, and conformism. Conformism is the phenomenon that persons tend to align opinions to the group opinion, especially when speaking in front of the group, which is also called groupthink [39]. Groupthink is a negative form of convergence, where confirmation is not rational but based on social pressure. These factors can be effectively dealt with by the use of anonymity in the electronic modeling tool [35]. A downside of anonymity is that changes cannot be attributed to experts or stakeholders.

4.3 Modeling Method

One of the main challenges of collaborative modeling is to design the process for the modeling effort, i.e. a sequence of modeling steps [6, 8]. In [8], Andersen and Richardson plea for a flexible approach, where the structure of each modeling effort is adapted to the context and may even be adapted during the sessions. Dean et al. examine modeling methods from a less flexible perspective [13].

A key question in the design of the collaborative modeling process is whether to start from scratch or with a preliminary model that is created by the supporting team based on interviews or documents [40]. Vennix reports that the use of a preliminary model is most suitable for cases where time is costly or where the supporting team is less experienced, because it can increase efficiency an encourage a lively discussion right from the start [8, 40]. On the other hand, there is a danger of perceived lack of ownership of the preliminary model which could result in low commitment, putting the process on the agenda, or rejection of the model [15, 40]. The process may be thwarted if the preliminary model is based on unknown assumptions or outdated [34].

When starting from scratch, several approaches are available. In most approaches the first step consists of a brainstorm or "gathering" to elicit the relevant concepts [12, 13, 35]. In approaches for Enterprise Analysis [13] an initial model is often created during the beginning of the modeling effort, often after a brainstorm about the most relevant high-level concepts. This initial model acts as a starting point for the further development, but differs from the preliminary models often used in Group Model Building [6, 8] in that it is built in-session with the plenary group rather than by the supporting team in advance. Furthermore, although both starting models can change considerably during the overall process (and it is stressed that they will), initial models (built with the group) will probably determine the structure of the final model more than preliminary models (built by the support team), because preliminary models aim to provoke an initial discussion to elicit conflicting assumptions and perspectives [13, 40]. The use of a preliminary model is extended in the so-called prototyping strategy, where for each step in the modeling process an analyst prepares the model and participants subsequently criticize and change the model [20]. Hengst and de Vreede write that this approach produces better results than when participants carry out each step, or when an analyst carries out each step [34].

In the subsequent convergence phase a different emphasis emerges among modeling approaches. Problem Structuring Methods have a stronger focus on eliciting the relations between individual mental maps as the model is considered a means to achieve consensus and shared understanding [12, 16]. The approaches for Enterprise Analysis focus more on the structure of the model itself, and are therefore more based on the grammar, and focus on correctness of individual modeling techniques, i.e. IDEF0-standards [13].

4.4 Model Quality

The importance of the model quality can differ for each case. For example, if the goal is to learn to improve collaboration and teamwork the modeling process is more important than its output [18], or when the goal is to learn the modeling method syntactic quality is critical [41]. Quality is a container concept for 'meeting criteria'. Depending on the goal of the collaborative modeling effort criteria for quality can be determined. In modeling, quality has two key classes of criteria: syntactic and a semantic quality. Syntactic quality concerns the correctness of the model according to the grammar of the modeling language, and therewith it's explanatory power. Semantic quality concerns the correctness of the model in terms of content, and whether it represents the system it describes. In collaboration quality is focused on process (e.g. participation, progress) and outcome (e.g. efficiency, effectiveness, complexity, shared understanding), and can be objective (e.g. time spend, quality according to experts) and subjective (e.g. satisfaction, usefulness).

4.4.1 Syntactic Quality

The syntactic quality of a model can be measured according to predefined model type-specific syntactic rules. These rules are most prescribed for the IDEF0 format, and most attention for syntax is found in the literature on the Enterprise Analysis. Apart from IDEF0-specific rules, model quality aspects that are important for all modeling types are the low amount of homonyms (some concepts are included in others) and synonyms (overlapping concepts), and the interconnectedness of different parts of the model, the latter being more a semantic qualifier. These aspects are especially important when the model is built by participants in a parallel setting [22, 23]. Dennis et al. found that, as would be expected, models that are built by an experienced modeler have better syntactical quality than models that are interactively built by participants [23, 24]. Therefore, approaches for parallel modeling with a high level of participation have to incorporate ways to improve syntactic quality. There are several methods at hand to this aim: training, guidance, periodic review, change awareness, and technological support. An extensive training of several days might be desired for the syntactic quality, but is often unpractical and costly. Therefore, Dean et al. suggest a combination of a small training and guidance of an experienced modeler during the sessions [13]. Model integration can also be improved through an explicit integration process step with the plenary group, whereas integration of participant built models by an external integrator can cause feelings of loss of ownership [23, 24]. Further, a modeling tool can have various change awareness functionalities, with which a subgroup can view the changes made by other subgroups or are automatically notified about these changes. Some case studies report good results where subgroups integrated their own model parts with others during the parallel process step [13, 42], but there is little evidence that supports the claim that a support team can rely on the voluntary integration by participants. Technological support can be used to avoid the appearance of homonyms and synonyms: before defining a new input a user has to go through a list of previously defined concepts. Also, built-in restrictions according to syntactic rules can improve syntactical quality, which proved very successful [13, 22].

4.4.2 Semantic Quality

In practice, semantic model quality can be difficult to measure, so one has to rely on the subjective perceptions of participants or the support team [40]. We note that in few studies perceptions of model quality are measured from a participants' perspective. Semantic model quality concerns the completeness and correctness of the model [34]. The completeness of a model denotes to what extent the model covers all aspects of the system it represents. A high complexity can be an indication of completeness , at the same time, models are meant to offer insight in an aspect or part of a system and should therewith reduce complexity [29]. There a several ways to measure complexity, either quantitatively by the number of objects and relations in the model, or qualitatively though observation, interviewing or analysis of results. The correctness of a model denotes to what extent the aspects of a system are depicted adequately [34]. Hengst and de Vreede write that stakeholder involvement can produce more complete and correct models [34]. This may be due to more richness and diversity of expertise in the group. Also, the role of a modeler in the session may be of influence to the model quality. However, Dennis et al. found hardly any difference in semantic model quality between collaborative and analyst-built models [13, 23].

5 Discussion and Conclusions

This paper presented an overview of four critical challenges in collaborative modeling, the related findings from cases in literature and the trade-offs involved. This overview offers an overview of challenges, which provide a basis for the development of new supporting methods and systems to overcome these challenges and to empower participants and facilitators in collaborative modeling.

Within the field of collaborative modeling various different approaches use well-developed sophisticated methods which are specific for the approach. Although these approaches are designed to apply to different settings, we feel that research in collaborative modeling would benefit greatly from focusing on the similarities between them.

We found a couple of key trade-offs that have to be considered for successful collaborative modeling efforts.

- A first trade-off can be found in the choices with respect to group composition. On one hand involving stakeholders and experts can improve correctness, buy-in and completeness; on the other hand it can lead to conflict and misunderstanding due to different perspectives and non-overlapping expertise. In smaller groups, model building efficiency will be higher, participation will increase and it is generally easier to create shared understanding.
- A second important trade-off was found with respect to the level of participation. If participants are empowered to make changes to the model directly, they will have a feeling of ownership and are more likely to accept the final model and decisions derived from it. However, critical stakeholders and domain experts are not necessarily skilled modelers. To achieve syntactical quality of the model it is therefore useful to involve a chauffeur or modeler. The trade-off between quality and participation has to be evaluated in light of the scope and complexity of the system that is to be represented. Further research has to be done to evaluate whether the role of modeler can be performed by participants themselves.
- A third critical challenge is the choice of a starting point for the modeling task. The use of a preliminary model, created by an expert or analyst, outside the group process, can speed up the process and raise critical discussion topics, but can also cause detachment and even rejection of the process and the resulting model.
- A final challenge can be found when collaborative modeling effort is performed in parallel, which can also improve modeling efficiency. When separate (sub) models are created in parallel a challenge lies in the convergence and integration of these models In order to support integration of sub models or changes created in parallel, strict rules are required to ensure syntactical quality and shared understanding. An interesting research challenge lies in the development and evaluation of facilitation tools and techniques to support the integration of sub-models. Such research can benefit from the use of patterns in access control and change awareness for computer-mediated interaction [43]. Another interesting direction for further research

is to find other convergence techniques to integrate different perspectives and to resolve conflicts in semantics, perceived relations, and scope.

Concluding, the field of collaborative modeling has a rich history, but researchers have only just begun to capture and formalize best practices and methods that are known to help achieve successful outcomes. Lessons learned are numerous, but the challenges identified in this paper offer a research agenda to develop formalized methods for collaborative modeling and to design new tools to support these methods.

References

1. Boehm, B., Gruenbacher, P., Briggs, R.O.: Developing Groupware for Requirements Negotiation: Lessons Learned. IEEE Software 18 (2001)
2. Fruhling, A., de Vreede, G.J.: Collaborative Usability Testing to Facilitate Stakeholder Involvement. In: Biffl, S., Aurum, A., Boehm, B., Erdogmus, H., Grünbacher, P. (eds.) Value Based Software Engineering, pp. 201–223. Springer, Berlin (2005)
3. Standish Group: CHAOS Report: Application Project and Failure (1995)
4. Hill, R.C., Levenhagen, M.: Methaphors and Mental Models: Sensemaking and Sensegiving in Innovative and Entrepreneurial Activities. Journal of Management 21, 1057–1074 (1995)
5. Frost, Sullivan.: Meetings Around the World: The Impact of Collaboration on Business Performance. Frost & Sullivan White Papers, 1–19 (2007)
6. Vennix, J.A.M., Andersen, D.F., Richardson, G.P., Rohrbaugh, J.: Model-building for group decision support: Issues and alternatives in knowledge elicitation support. European Journal of Operational Research 59, 28–41 (1992)
7. Rouwette, E.A.J.A., Vennix, J.A.M., Mullekom, T.v.: Group Model Building Effectiveness: a Review of Assessment Studies. System Dynamics Review 18, 5–45 (2002)
8. Andersen, D.F., Richardson, G.P.: Scripts for Group Model Building. System Dynamics Review 13, 107–129 (1997)
9. Richardson, G.P., Andersen, D.F.: Teamwork in Group Model Building. System Dynamics Review 11, 113–137 (1995)
10. Mulder, I., Swaak, J., Kessels, J.: Assessing learning and shared understanding in technology-mediated interaction. Educational Technology & Society 5, 35–47 (2002)
11. Weick, K.E.: Sensemaking in Organizations. Sage Publications Inc., Thousand Oaks (1995)
12. Shaw, D., Ackermann, F., Eden, C.: Approaches to sharing knowledge in group problem structuring. Journal of the Operational Research Society 54(913), 936–948 (2003)
13. Dean, D.L., Orwig, R.E., Vogel, D.R.: Facilitation Methods for Collaborative Modeling Tools. Group Decision and Negotiation 9, 109–127 (2000)
14. Rosenhead, J.: Rational analysis for a problematic world: problem structuring methods for complexity, uncertainty and conflict (1993)
15. Eden, C., Ackermann, F.: Cognitive mapping expert views for policy analysis in the public sector. European Journal of Operational Research 127, 615–630 (2004)
16. Eden, C., Ackermann, F.: Where next for Problem Structuring Methods. Journal of the Operational Research Society 57, 766–768 (2006)
17. Ackermann, F., Eden, C.: Using Causal Mapping with Group Support Systems to Elicit an Understanding of Failure in Complex Projects: Some Implications for Organizational Research. Group Decision and Negotiation 14, 355–376 (2005)

18. Ackermann, F., Franco, L.A., Gallupe, B., Parent, M.: GSS for Multi-Organizational Collaboration: Reflections on Process and Content. Group Decision and Negotiation 14, 307–331 (2005)
19. Andersen, D.F., Vennix, J.A.M., Richardson, G.P., Rouwette, E.A.J.A.: Group model building: problem structuring, policy simulation and decision support. Journal of the Operational Research Society 58, 691–694 (2007)
20. Lund, J.R., Palmer, R.N.: Water Resource System Modeling for Conflict Resolution. Water Resources Update 108, 70–82 (1997)
21. Morton, A., Ackermann, F., Belton, V.: Technology-driven and model-driven approaches to group decision support: focus, research philosophy, and key concepts. European Journal of Information Systems 12, 110–126 (2003)
22. Dean, D.L., Lee, J.D., Orwig, R.E., Vogel, D.R.: Technological Support for Group Process Modeling. Journal of Management Information Systems 11, 43–63 (1994)
23. Dennis, A.R., Hayes, G.S., Daniels Jr., R.M.: Re-engineering Business Process Modeling. In: Proceedings of the Twenty-Seventh Annual Hawaii International Conference on System Sciences (1994)
24. Dennis, A.R., Hayes, G.S., Daniels Jr., R.M.: Business process modeling with group support systems. Journal of Management Information Systems 15, 115–142 (1999)
25. Gurung, T.R., Bousquet, F., Trébuil, G.: Companion Modeling, Conflict Resolution, and Institution Building: Sharing Irrigation Water in the Lingmuteychu Watershed, Bhutan. Ecology and Society 11 (2006)
26. Trauth, E.M., Jessup, L.M.: Understanding Computer-mediated Discussions: Positivist and Interpretive Analyses of Group Support System Use. MIS Quarterly 24, 43–79 (2000)
27. Creswell, J.W.: Research Design: Qualitative & Quantitative Approaches. Sage Publications, Inc., Thousand Oaks (1994)
28. Webster, J., Watson, R.T.: Analyzing the Past to Prepare for the Future: Writing a Literature Review. MIS Quarterly 26, xiii–xxiii (2002)
29. Nunamaker, J.F., Alan, R.D., Joseph, S.V., Douglas, V., Joey, F.G.: Electronic meeting systems to support group work, vol. 34, pp. 40–61. ACM, New York (1991)
30. Kolfschoten, G.L., Niederman, F., Vreede, G.J.d., Briggs, R.O.: Roles in Collaboration Support and the Effect on Sustained Collaboration Support. In: Hawaii International Conference on System Science. IEEE Computer Society Press, Waikoloa (2008)
31. Vreede, G.J.d., Boonstra, J., Niederman, F.A.: What is Effective GSS Facilitation? A Qualitative Inquiry into Participants' Perceptions. In: Hawaiian International Conference on System Science. IEEE Computer Society Press, Los Alamitos (2002)
32. Dennis, A.R., Wixom, B.H., Vandenberg, R.J.: Understanding Fit and Appropriation Effects in Group Support Systems Via Meta-Analysis. Management Information Systems Quarterly 25, 167–183 (2001)
33. Griffith, T.L., Fuller, M.A., Northcraft, G.B.: Facilitator Influence in Group Support Systems. Information Systems Research 9, 20–36 (1998)
34. den Hengst, M., de Vreede, G.J.: Collaborate Business Process Engineering: A Decade of Lessons from the Field. Journal of Management Information Systems 20, 85–113 (2004)
35. Rouwette, E.A.J.A., Vennix, J.A.M., Thijssen, C.M.: Group Model Building: A Decision Room Approach. Simulation & Gaming 31, 359–379 (2000)
36. van den Belt, M.: Mediated Modeling: A System Dynamics Approach to Environmental Consensus Building. Island Press (2004)
37. Vreede, G.J.d., Davison, R., Briggs, R.O.: How a Silver Bullet May Lose its Shine - Learning from Failures with Group Support Systems. Communications of the ACM 46, 96–101 (2003)

38. Maghnouji, R., de Vreede, G., Verbraeck, A., Sol, H.: Collaborative Simulation Modeling: Experiences and Lessons Learned. In: HICSS 2001: Proceedings of the 34th Annual Hawaii International Conference on System Sciences (HICSS-34), vol. 1, p. 1013. IEEE Computer Society, Washington (2001)
39. Janis, I.L.: Victims of Groupthink: A Psychological Study of Foreign-Policy Decisions and Fiascoes. Houghton Mifflin Company, Boston (1972)
40. Vennix, J.A.M.: Group Model Building: Facilitating Team Learning Using System Dynamics. John Wiley & sons, Chichester (1996)
41. Hengst, M.d.: Collaborative Modeling of Processes: What Facilitation Support does a Group Need? In: Americas Conference on Information Systems, AIS Press, Omaha (2005)
42. Ram, S., Ramesh, V.: Collaborative conceptual schema design: a process model and prototype system. ACM Transactions on Information Systems 16, 347–371 (1998)
43. Schümmer, T., Lukosch, S.: Patterns for Computer-Mediated Interaction. Wiley & Sons Ltd., West Sussex (2007)

Appendix: Articles Used in the Literature Review

Ackermann, F., Eden, C.: Using Causal Mapping with Group Support Systems to Elicit an Understanding of Failure in Complex Projects: Some Implications for Organizational Research. Group Decision and Negotiation, Vol . 14 (2005) 355-376

Ackermann, F., Franco, L., Gallupe, B., Parent, M.: GSS for Multi-Organizational Collaboration: Reflections on Process and Content. Group Decision and Negotiation **14** (2005) 307-331

Adamides, E.D., Karacapilidis, N.: A Knowledge Centred Framework for Collaborative Business Process Modelling. Business Process Management, Vol. 12 (2006) 557-575

Akkermans, H.A., Vennix, J.A.M.: Clients' Opinions on Group Model-Building: An Exploratory Study. System Dynamics Review, Vol. 13 (1997) 3-31

Andersen, D.F., Richardson, G.P.: Scripts for Group Model Building. System Dynamics Review, Vol. 13 (1997) 107-129

Andersen, D.F., Vennix, J.A.M., Richardson, G.P., Rouwette, E.A.J.A.: Group Model Building: Problem Structuring, Policy Simulation and Decision Support. Journal of the Operational Research Society, Vol. 58 (2007) 691-694

Aytes, K.: Comparing Collaborative Drawing Tools and Whiteboards: An Analysis of the Group Process. Computer Supported Cooperative Work, Vol. 4. Kluwer Academic Publishers (1995) 51-71

van den Belt, M.: Mediated Modeling: A System Dynamics Approach to Environmental Consensus Building. Island Press (2004)

de Cesare, S., Serrano, A.: Collaborative Modeling Using UML and Business Process Simulation. HICSS '06: Proceedings of the 39th Annual Hawaii International Conference on System Sciences. IEEE Computer Society, Washington, DC, USA (2006) 10.12

Daniell, K.A., Ferrand, N., Tsoukia, A.: Investigating Participatory Modelling Processes for Group Decision Aiding in Water Planning and Management. Group Decision and Negotiation (2006)

Dean, D., Orwig, R., Lee, J., Vogel, D.: Modeling with a Group Modeling Tool: Group Support, Model Quality and Validation. Proceedings of the Twenty-Seventh Annual Hawaii International Conference on System Sciences (1994)

Dean, D.L., Lee, J.D., Nunamaker, J.J.F.: Group Tools and Methods to Support Data Model Development, Standardization, and Review. Proceedings of the Thirtieth Annual Hawaii International Conference on System Sciences (1997)

Dean, D.L., Lee, J.D., Orwig, R.E., Vogel, D.R.: Technological Support for Group Process Modeling. Journal of Management Information Systems, Vol. 11. M. E. Sharpe, Inc. (1994) 43-63

Dean, D.L., Lee, J.D., Pendergast, M.O., Hickey, A.M., Jay F. Nunamaker, Jr.: Enabling the Effective Involvement of Multiple Users: Methods and Tools for Collaborative Software Engineering. Journal of Management Information Systems, Vol. 14. M. E. Sharpe, Inc. (1997) 179-222

Dean, D.L., Orwig, R.E., Vogel, D.R.: Facilitation Methods for Use with EMS Tools to Enable Rapid Development of High Quality Business Process Models. Proceedings of the 29th Hawaii International Conference on System Sciences, Vol. 3. IEEE Computer Society, Washington, DC, USA (1996) 472

Dean, D.L., Orwig, R.E., Vogel, D.R.: Facilitation Methods for Collaborative Modeling Tools. Group Decision and Negotiation, Vol. 9 (2000) 109-127

Dennis, A.R., Hayes, G.S., Daniels, R.M., Jr.: Re-engineering Business Process Modeling. Proceedings of the Twenty-Seventh Annual Hawaii International Conference on System Sciences (1994)

Dennis, A.R., Hayes, G.S., Robert M. Daniels, Jr.: Business Process Modeling with Group Support Systems. Journal of Management Information Systems, Vol. 15. M. E. Sharpe, Inc. (1999) 115-142

Eden, C., Ackermann, F.: Cognitive Mapping Expert Views for Policy Analysis in the Public Sector. European Journal of Operational Research **127** (2004) 615-630

Eden, C., Ackermann, F.: Where Next for Problem Structuring Methods. Journal of the Operational Research Society, Vol. 57 (2006) 766-768

Gurung, T.R., Bousquet, F., Trébuil, G.: Companion Modeling, Conflict Resolution, and Institution Building: Sharing Irrigation Water in the Lingmuteychu Watershed, Bhutan. Ecology and Society, Vol. 11 (2006)

Hayne, S., Ram, S.: Group Data Base design: Addressing the View Modeling Problem. J. Syst. Softw., Vol. 28. Elsevier Science Inc. (1995) 97-116

den Hengst, M., de Vreede, G.J.: Collaborate Business Process Engineering: A Decade of Lessons from the Field. Journal of Management Information Systems, Vol. 20 (2004) 85-113

den Hengst, M., de Vreede, G.J., Magnhnouji, R.: Using soft OR Principles for Collaborative Simulation: A Case Study in the Dutch Airline Industry. Journal of the Operational Research Society, Vol. 58 (2006) 669-682

Jefferey, A.B., Maes, J.D.: Improving Team Decision-Making Performance with Collaborative Modeling. Team Performance Management, Vol. 11 (2005) 40-50

Lee, J.D., Dean, D.L., Vogel, D.R.: Tools and Methods for Group Data Modeling: A Key Enabler of Enterprise Modeling. SIGGROUP Bulletin, Vol. 18. ACM (1997) 59-63

Lee, J.D., Zhang, D., Santanen, E., Zhou, L., Hickey, A.M.: ColD SPA: A Tool For Collaborative Process Model Development. HICSS '00: Proceedings of the 33rd Hawaii International Conference on System Sciences-Volume 1. IEEE Computer Society, Washington, DC, USA (2000) 1004

Lucia, A.D., Fasano, F., Scanniello, G., Tortora, G.: Enhancing Collaborative Synchronous UML Modelling with Fine-grained Versioning of Software Artefacts. J. Vis. Lang. Comput., Vol. 18. Academic Press, Inc. (2007) 492-503

Lund, J.R., Palmer, R.N.: Water Resource System Modeling for Conflict Resolution. Water Resources Update, Vol. 108 (1997) 70-82

Maghnouji, R., de Vreede, G., Verbraeck, A., Sol, H.: Collaborative Simulation Modeling: Experiences and Lessons Learned. HICSS '01: Proceedings of the 34th Annual Hawaii International Conference on System Sciences (HICSS-34)-Volume 1. IEEE Computer Society, Washington, DC, USA (2001) 1013

Millward, S.M.: Do You Know Your STUFF? Training Collaborative Modelers. Team Performance Management, Vol. 12 (2006) 225-236

Orwig, R., Dean, D.: A Method for Building a Referent Business Activity Model for Evaluating Information Systems: Results from a Case Study. Communications of the AIS **20** (2007) article 53

Pata, K., Sarapuu, T.: A Comparison of Reasoning Processes in a Collaborative Modelling Environment: LearningAbout Genetics Problems Using Virtual Chat. International Journal of Science Education, Vol. 28 (2006) 1347-1368

Purnomo, H., Yasmi, Y., Prabhu, R., Hakim, S., Jafar, A., Suprihatin: Collaborative Modelling to Support Forest Management: Qualitative Systems Management at Lumut Mountain, Indonesia. Small Scale Forest Economics, Management and Policy, Vol. 2 (2003) 259-275

Ram, S., Ramesh, V.: Collaborative Conceptual Schema Design: A Process Model and Prototype System. ACM Transactions on Information Systems, Vol. 16. ACM (1998) 347-371

Richardson, G.P., Andersen, D.F.: Teamwork in Group Model Building. System Dynamics Review, Vol. 11 (1995) 113-137

Rouwette, E.A.J.A., Vennix, J.A.M., Thijssen, C.M.: Group Model Building: A Decision Room Approach. Simulation & Gaming, Vol. 31 (2000) 359-379

Rouwette, E.A.J.A., Vennix, J.A.M., van Mullekom, T.: Group Model Building Effectiveness: A Review of Assessment Studies. System Dynamics Review, Vol. 18 (2002) 5-45

Samarasan, D.: Collaborative Modeling and Negotiation. SIGOIS Bulletin, Vol. 9. ACM (1988) 9-21

Vennix, J.A.M.: Group Model Building: Facilitating Team Learning Using System Dynamics. John Wiley & sons (1996)

Vennix, J.A.M.: Group Model-Building: Tackling Messy Problems. System Dynamics Review, Vol. 15 (1999) 379-401

Vennix, J.A.M., Akkermans, H.A., Rouwette, E.A.J.A.: Group Model-building to Facilitate Organizational Change: An Exploratory Study. System Dynamics Review, Vol. 12 (1996) 39-58

Vennix, J.A.M., Andersen, D.F., Richardson, G.P.: Foreword: Group Model Building, Art, and Science. System Dynamics Review, Vol. 13 (1997) 103-106

Vennix, J.A.M., Andersen, D.F., Richardson, G.P., Rohrbaugh, J.: Model-building for Group Decision Support: Issues and Alternatives in Knowledge Elicitation Support. European Journal of Operational Research **59** (1992) 28-41

de Vreede, G.-J.: Facilitating Organizational Change. Delft University of Technology (1995)

de Vreede, G.-J.: Group Modeling for Understanding. Journal of Decision Systems, Vol. 6 (1997) 197-220

A Petri-Net Based Formalisation of Interaction Protocols Applied to Business Process Integration

Djamel Benmerzoug[1], Fabrice Kordon[2], and Mahmoud Boufaida[1]

[1] LIRE Laboratory, Computer Science Department,
Mentouri University of Constantine 25000, Algeria
{benmerzougdj,boufaida_mahmoud}@yahoo.fr
[2] LIP6 Laboratory, Pierre et Marie Curie University,
4, place Jussieu, 75252 Paris Cedex 05 France
fabrice.kordon@lip6.fr

Abstract. This paper presents a new approach for Business Process Integration based on Interaction Protocols. It enables both integration and collaboration of autonomous and distributed business processes modules. We present a semantic formalisation of the interaction protocols notations used in our approach. The semantics and its application are described on the basis of translation rules to Coloured Petri Nets and the benefits of formalisation are shown. The verified and validated interaction protocols specification is exploited afterwards with an intermediate agent called « Integrator Agent » to enact the integration process and to manage it efficiently in all steps of composition and monitoring.

Keywords: Business Processes Integration, Interaction Protocols, Coloured Petri Nets, Multi-agent Systems.

1 Introduction

Unlike traditional business processes, processes in open, Web-based settings typically involve complex interactions among autonomous, heterogeneous business partners. In such environments there is a clear need for advanced business applications to coordinate multiple business processes into a multi-step business transaction. This requires that several business operations or processes attain transactional properties reflecting business semantics, which are to be treated as a single logical unit of work [1]. This orientation requires distilling from the structure of businesses collaboration the key capabilities that must necessarily be present in a Business Process Integration (BPI) scenario and specifying them accurately and independently from any specific implementation mechanisms.

Web services are a promising technology to support business processes coordination and collaboration [2][3]. They are an XML-based middleware that provides RPC-like remote communication, using in most cases SOAP over HTTP. Web services are designed to allow machine-to-machine interactions. This interaction takes place over a network, such as the Internet, so Web services are by definition distributed, and operate in an open and highly dynamic environment.

Heterogeneity, distribution, openness, highly dynamic interactions, are some among the key characteristics of another emerging technology, that of intelligent agents and

J.L.G. Dietz et al. (Eds.): CIAO! 2008 and EOMAS 2008, LNBIP 10, pp. 78–92, 2008.

Multi-Agent Systems (MAS). M. Luck et al. [4] propose the following definition: "an agent is a computer system that is capable of flexible autonomous action in dynamic, unpredictable, typically multi-agent domains."

We already proposed a new approach based on Web services and agents for integrating business processes [5]. The BPI modeling is based on Interaction Protocols (IP) that enable autonomous, distributed business process management modules to integrate and collaborate.

IP are a useful way for structuring communicative interaction among business process management modules, by organizing messages into relevant contexts and providing a common guide to all parties. The value of IP-based approach is largely determined by the interaction model it uses. The presence of an underlying formal model supports the use of structured design techniques and formal analysis and verification, facilitating development, composition and reuse.

Most IP modeling projects to date have used or extended finite state machines (FSM) and state transition diagram (STD) in various ways [8]. FSM and STD are simple, depict the flow of action/communication in an intuitive way, and are sufficient for many sequential of interactions. However, they are note adequately expressive to model more complex interactions, especially those with some degree of concurrency. In the other hand, Coloured Petri Nets (CPN) [9] are a well known and established model of concurrency, and can support the expression of a greater range of interactions. In addition, CPN like FSM, have an intuitive graphical representation, are relatively simple to implement, and are accompanied with a variety of techniques and tools for formal analysis and design.

Unfortunately, the existing works on the use of formal models to represent IP leave open several questions [8], [16], [19], [21]. Most previous investigations have not provided a systematic comprehensive coverage of all issues that arise when representing complex protocols such as intra-Enterprise Application Integration (EAI) as well as the inter-enterprise integration (B2B, for Business to Business).

This paper presents a generic approach for the BPI based on interaction protocols. Translation rules of IP based on AUML/BPEL4WS [13],[14] notations into CPN are proposed, enabling their formal analysis and verification. We provide interactions building blocks allowing this translation to model complex e-business applications that enable autonomous, distributed business process management modules to integrate and collaborate.

This CPN-based representation can be used to essentially cover all the features used in IP standards, including communicative act attributes (such as message guards and cardinalities) and protocol nesting. Also, we present a skeleton automated procedure for converting an IP specification to an equivalent CPN, and demonstrate its use through a case study.

In the next section we, briefly present our approach. Section 3 describes a CPN based representation of IP. In section 4, we provide a skeletal algorithm for converting BPI based on interaction protocols in AUML/BPEL4WS to Coloured Petri nets. Section 5 shows how the verified and the validated IP specification can be exploited by the MAS to enact the BPI. Related work is discussed in section 6 and conclusions are drawn in section 7.

2 An Overview of the Proposed Approach

In recent years, BPI modeling and reengineering have been longstanding activities in many companies. Most internal processes have been streamlined and optimized, whereas the external processes have only recently become the focus of business analysts and IT middleware providers. The static integration of inter-enterprise processes as common in past years can no longer meet the new requirements of customer orientation, flexibility and dynamics of cooperation [10].

In [6],[7] we have developed an agent-based method for developing cooperative enterprises information systems. This method permits to explicitly map the business process into software agents. In [5], we have described the use of IP to define and manage public processes in B2B relationships. This process is modelled using AUML (Agent UML [13]) and specified with BPEL4WS [14].

In this approach, we consider two types of business processes, the private processes and the public ones. The first type is considered as the set of processes of the company itself and they are managed in an autonomous way. Private processes are supported within companies using traditional Workflow Management Systems, Enterprise Resources Planning systems or proprietary systems. These systems were intended to serve local needs. In other hand, public processes span organizational boundaries. They belong to the companies involved in a B2B relationship and have to be agreed and jointly managed by the partners.

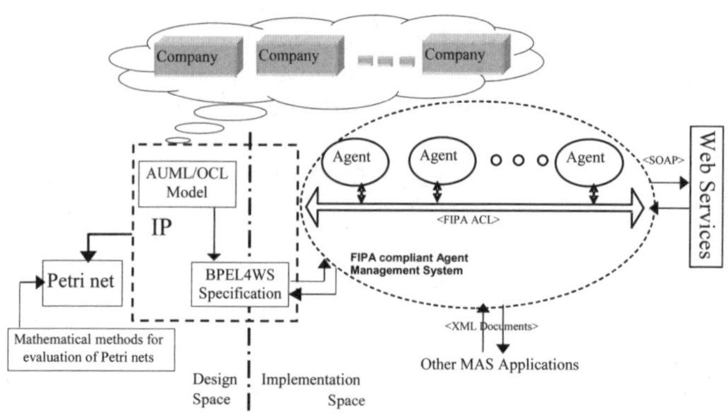

Fig. 1. The proposed approach

The B2B integration scenarios typically involve distributed business processes that are autonomous to some degree. Companies participating in this scenario publish and implement a public process. The applications integration based on public process is not a new approach. The current models for BPI are based on process flow graphs [11], [12]. A process flow graph is used to represent the public process. This approach lacks the flexibility for supporting dynamic B2B integration. In contrast, our approach (figure 1) presents an incremental, open-ended, dynamic, and personalizable model for B2B integration.

The use of IP to define public processes enables a greater autonomy of companies because each company hides its internal activities, services and decisions required to support public processes. In this way, the IP provide a high abstraction level in the modelling of public processes. The AUML model is mapped to a BPEL4WS specification, which represents the initial social order upon a collection of agents (figure 1). Since BPEL4WS describes the relationships between the Web services in the public process, agents representing the Web services would know their relationships a priori. Notably, the relationships between the Web services in the public process are embedded in the process logic of the BPEL4WS specification.

This relationship entails consistency problems, which can at best be solved at the level of models. Indeed, we used the BPEL4WS specification to generate a validation tool that can check that a BPEL4WS document is well-formed (the BPEL4WS preserves the business constraints, which are specified by means of OCL (Object Constraint Language [23])). In this work, we have exploited the Sun Microsystem Web Services Developer Pack [15]. In particular, we have used the JAXB (Java Architecture for XML Binding) library to build Java classes from a BPEL4WS specification (for more detail see [5]).

In this paper, we address the problem of verification of BPI based on interaction protocols. Indeed, we propose a novel and flexible representation of protocols that uses CPN in which, interaction building blocks explicitly denote joint conversation states and messages. So, interaction protocols specification can be translated to an equivalent CPN model and CPN tools can afterwards be used to analyze the process.

3 A CPN-Based Model for BPI Based on Interaction Protocol

BPI is defined as an interaction protocol involving different companies. It specifies the interaction between local business process and Web services and their coordination. For this purpose, we define the IP as follow:

Definition: An Interaction Protocol is a quadruplet: IP = <ID, R, M, f_M>, where:

- ID is the identify of the interaction protocol
- R = $\{r_1, r_2, \ldots, r_n\}$ (n>1) is a set of Roles (private business process or Web Services)
- M is a set of non-empty primitive (or/and) complex messages, where:
    - A Primitive Message (PM) corresponds to the simple message, it is defined as follow: PM = <Sender, Receiver, CA, Option>, where:
        - Sender, Receiver $\in$ R
        - CA $\in$ FIPA ACL (**C**ommunicative **A**ct such as: cfp, inform, …)
        - Option: contain additional information (Synchronous / Asynchronous message, constraints on message, …)
    - A Complex Message (CM) is built from simpler (primitive) ones by means of operators: CM = PM_1 op PM_2 … op PM_m.where:
        - m>1, op $\in$ {OR, XOR, AND}, and
        - $\forall$ i $\in$ [1, m[, PM_i.Sender = PM_{i+1}.Sender, PM_i.Sender $\in$ R .
- f_M: a flow relation defined as : $f_M \subseteq$ (RxR), where (RxR) is a Cartesian product (r1,r2) $\in$ (RxR), for r1,r2 $\in$ R

Ideally, IP should be represented in a way that allows performance analysis, validation and verification, automated monitoring, debugging, etc. Various formalisms have been proposed for such purposes. However, Petri nets have been shown to offer significant advantages in representing IP, compared to other approaches [16]. Specifically, Petri nets are useful in validation and testing, automated debugging and monitoring and dynamic interpretation of IP.

Our main motivation in describing the semantics of IP applied to BPI by using CPN is that the existence of several variation points allows different semantic interpretations that might be required in different application domains. This is usually our case, and so, high-level Petri nets are used as formal specification. This provides the following advantages:

- CPN provide true concurrency semantics by means of the step concept, i.e. when at least two non-conflictive transitions may occur at the same time. It is the ideal situation for our application domain (several activities moving within the same space of states: the <flow> section in BPEL4WS).
- The combination of states, activities, decisions, primitives and complex message exchanges (namely fork-join constructions) means that the IP notations are very rich. CPN allow us to express, in the same formalism, both the kind of system we are dealing with and its execution.
- Formal semantic is better in order to carry out a complete and highly automated analysis for the system being designed.

3.1 Translation Rules from IP Elements to CPN

The objective of this section is to propose some general rules which may be applied to formally specify interaction protocols endowing them with a formal semantics. Such a semantics will enable the designer to validate his/her specifications. As shown in the translation rules in Table 1, we focus on the description of dynamic aspects of protocols using the CPN's elements (places, transitions, arcs, functions, variables and domains).

The CPN representation in Table 1 introduces the use of token colours to represent additional information about business processes interaction states and communicative acts of the corresponding interaction. The token colour sets are defined in the net declaration as follow: (the syntax follows standard CPN-notation [9])

```
Colour sets :
Communicative Act = with inform|cfp|propose|… ;
Role = string with "a".. "z" ; // Role = {r1, r2, …}, ri ∈ R
Content = string with "a".. "z" ;
Bool = with true|false;
MSG = record s,r: Role; CA: Communicative Act; C: Content
Variables:msg, msg1, msg2: MSG;        x: Bool;
```

The *MSG* colour set describes communicative acts interaction and is associated with the net message places. The *MSG's* coloured token is a record <s,r,ca,c>, where the *s* and *r* elements determine the sender and the receiver of the corresponding message. This elements have the colour set *ROLE*, which is used to identify business processes or/and Web services participating in the corresponding interaction. The *COMMUNICATIVE ACT* and the *CONTENT* colour sets represent respectively the FIPA-ACL communicative acts and the content of the corresponding message. We note that places without colour set hold an indistinguishable token and therefore have the colour domain token = {●}.

We now show how various interaction protocols features described in our work can be represented using the CPN formalism.

R1: *A role* (the <partner> section in BPEL4WS) is considered equivalent to a type of resource, which is represented in a Petri net as a place. Hence, there will be one token in the place for each actor playing this role. Each one of these places is labelled with the corresponding role name.

R2: *The "life line" of role* is represented implicitly by a places and transitions sequence belonging to this role. The net is constituted therefore by one sub-net (Petri net process) for each role acting during the interaction and these nets are connected by places that correspond to the exchanged messages.

R3: *A message exchange* between two roles is represented by a synchronization place and arcs. The first ongoing arc connects the transition of "message sending" to the "synchronization place", while the second outgoing arc connects this place to the "receiving message transition".

R4: *A primitive message exchange:* As we have already said, a primitive message corresponds to the simple message. A <receive> and <reply> activities (asynchronous messages) are represented by a transition which has an in-place and out-place (see R3 in Table 1). An <invoke> activity (synchronous messages) is represented by a pair of transitions, one of them may fire a request token to the sub-net of the receiver role, and the other may wait for a token from this sub-net.

R5: *A complex message exchange:* A complex message is represented by a substitution transition. The control flow between messages exchange is captured by connecting the activity-related transitions with arcs, places, and transitions purely used for control flow purpose. More refined control flow can be expressed using arc inscriptions and transition guard expressions.

Table 1 (R5 – (1)) shows a more complex interaction, called XOR-decision. (the <if>/<pick> section in the BPEL4WS specification) so that only one communicative act can be sent. In this case, each type of message is associated to a transition with a function on its input arc. The function plays the role of a filter, i.e. it control the firing of the transition corresponding to the message type. Table 1 (R5 – (2)) shows another complex interaction, the OR-parallel interaction (the <switch> section), in which the sender can send zero, one or more communicative acts (inclusively) to the designated recipients simulating an inclusive-or.

The last type of complex message is the *AND*-parallel (the <follow> section) which models concurrency messages sending. This type of complex interaction is represented by means of parallel case or multi-threading in CPN.

R6: *Iteration:* An iteration in a part of IP specification is represented by an arrow and a guard expression or an end condition (the <while> section in BPEL4WS). In CPN, an iteration is specified in the same way except that the end condition is a guard expression associated with the transition that starts the iteration.

Table 1. A Translation Rules From IP to CPN

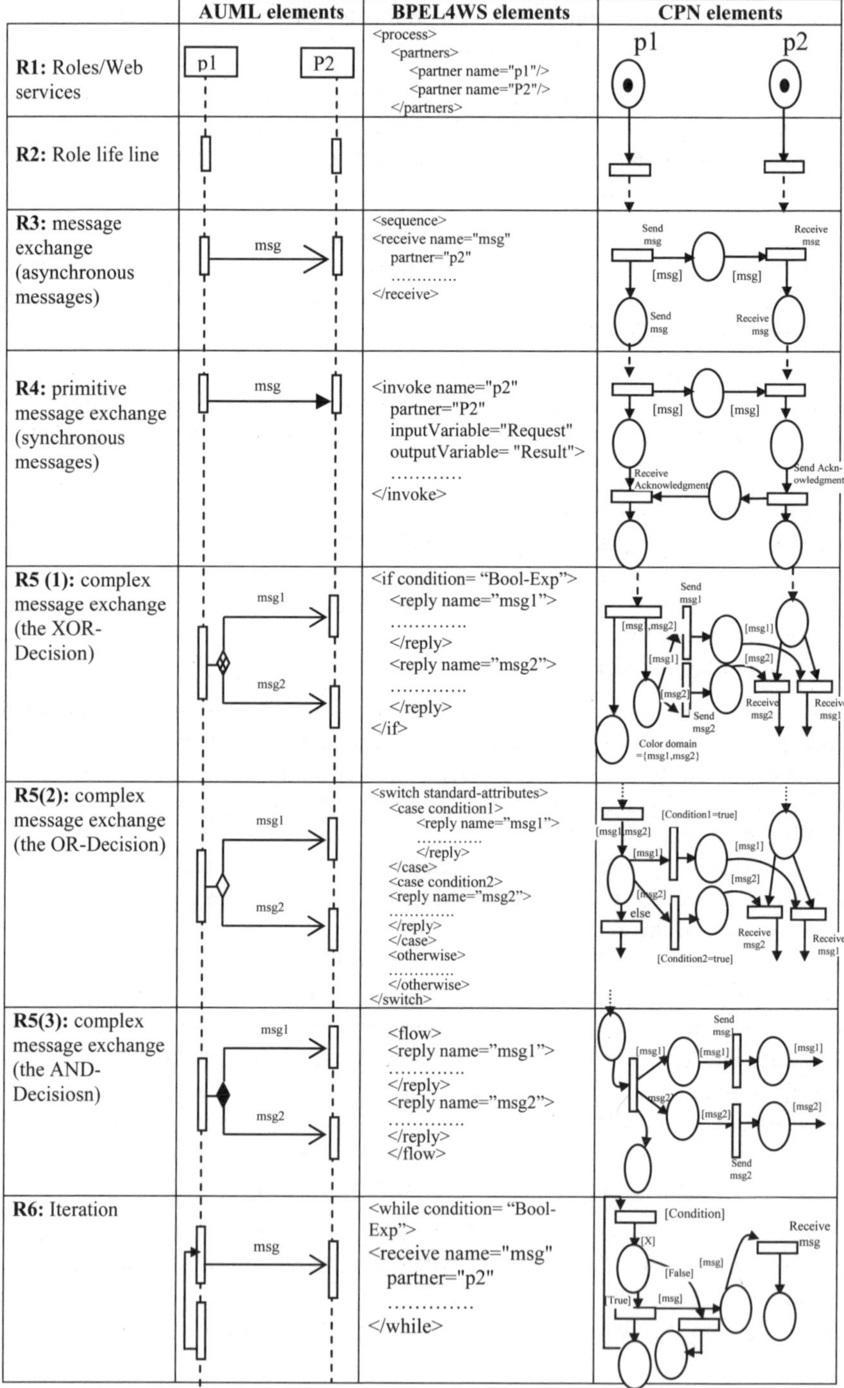

	AUML elements	BPEL4WS elements	CPN elements
R1: Roles/Web services	p1 P2	<process> <partners> <partner name="p1"/> <partner name="P2"/> </partners>	p1 p2
R2: Role life line			
R3: message exchange (asynchronous messages)	msg	<sequence> <receive name="msg" partner="p2" .... </receive>	Send msg Receive msg [msg] [msg] Send msg Receive msg
R4: primitive message exchange (synchronous messages)	msg	<invoke name="p2" partner="P2" inputVariable="Request" outputVariable= "Result"> .... </invoke>	[msg] [msg] Receive Acknowledgment Send Acknowledgment
R5 (1): complex message exchange (the XOR-Decision)	msg1 msg2	<if condition= "Bool-Exp"> <reply name="msg1"> .... </reply> <reply name="msg2"> .... </reply> </if>	Send msg1 [msg1,msg2] [msg1] [msg1] [msg2] [msg2] Send msg2 Receive msg2 Receive msg1 Color domain ={msg1,msg2}
R5(2): complex message exchange (the OR-Decision)	msg1 msg2	<switch standard-attributes> <case condition1> <reply name="msg1"> .... </reply> </case> <case condition2> <reply name="msg2"> .... </reply> </case> <otherwise> .... </otherwise> </switch>	[Condition1=true] [msg1,msg2] [msg1] [msg1] [msg2] [msg2] else [Condition2=true] Receive msg2 Receive msg1
R5(3): complex message exchange (the AND-Decisiosn)	msg1 msg2	<flow> <reply name="msg1"> .... </reply> <reply name="msg2"> .... </reply> </flow>	Send msg1 [msg1] [msg1] [msg1] [msg2] [msg2] [msg2] Send msg2
R6: Iteration	msg	<while condition= "Bool-Exp"> <receive name="msg" partner="p2" .... </while>	[Condition] [X] [False] [True] [msg] [msg] Receive msg

Table 1. *(continued)*

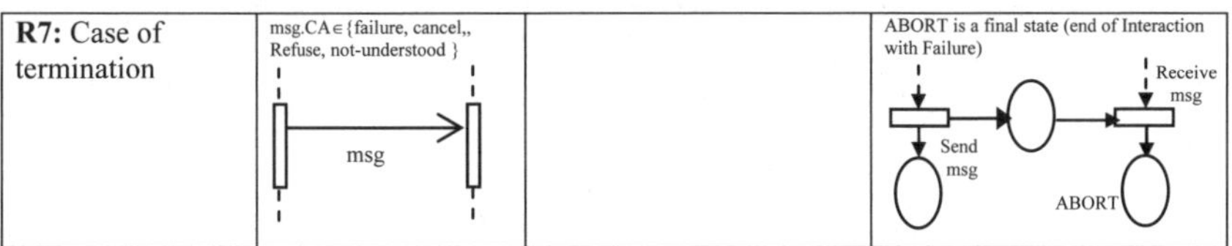

R7: Case of termination	msg.CA∈{failure, cancel,, Refuse, not-understood } msg		ABORT is a final state (end of Interaction with Failure) Receive msg Send msg ABORT

***R7:** Case of termination:* In the specification of the FIPA-ContractNetProtocol besides the AUML diagram other requirements are described in the text [13]: The sending of not-understood messages and the so called FIPA-Cancel-Metaprotocol: Every received message is responded to by a not-understood, if the comprehension of the message failed. In this case, the protocol is cancelled for the corresponding participant. In a CPN, this is realized by adding a transition to the final state ABORT (except the initial state). This transition corresponds to the reception of acts: Failure, Cancel, Refuse or not-understood, which can terminate the IP with failure.

3.2 An Algorithm for Transforming an IP to Its CPN Representation

Previous investigations have explored various machine-readable Petri net representations. However, interaction protocols are typically specified in human-readable form (e.g., in AUML [13]). The question of how to automatically translate an interaction protocol specification into a machine-readable form has been previously ignored [16]. We present an automated procedure for transforming an IP to its CPN representation.

The algorithm is presented in figure 2. It inputs an IP as defined in section 3, and it outputs a corresponding CPN representation. The CPN is constructed by iterating: The algorithm essentially creates the IP-net by exploring the interaction protocol. Lines 1 and 2 initiate different variables used in this algorithm and respectively the CPN output. The roles places, denoted by the variable RP, hold the initiating places for the Petri net. These places correspond to the roles of the IP (line 3, 4 and 5). Each one of these places is labelled with the corresponding role name.

We enter the main loop in line 7 and set *curr* to the first message in the IP. Lines 8-16 create the CPN components of the current iteration. First, in line 8, message places, associated with *curr* role place, are created using *CreateMessagePlace*.

These places correspond to communicative acts. Then, in line 9, we create intermediate places that correspond to interaction state changes as a result of these messages associated with *curr* place. Then, in *CreateTransitions* and *CreateArcs*, these places are connected through transitions and arcs, using the CPN building blocks previously described (section 3). Finally, we add token elements colour to the CPN structure, implementing attributes using the *FixColor* function (line 16).

To complete the iteration, the CPN output, is updated according to the current iteration in lines 17-19. The loop iterates as long as *M* contains messages that have not been handled. Finally, the resulting CPN is returned (line 21).

```
Algorithm CreateIP-net (input : IP=<ID, R, M, f_M>, output : CPN)

1: RP←Ø  // Roles places
   MP← Ø  // Messages places
   IM← Ø  // Intermediate places
   TR← Ø  // list of transitions
   AR← Ø  // list of arcs
2: CPN ← new CPN
3: For every r ∈ R do
4.    RP←createRolePlace() // there would be one token in every RP place
5: CPN.places ← RP
6: While M ≠ Ø do
7:   curr ← M.dequeue()
8:   MP← CreateMessagePlace(curr)
9:   IM ← CreateIntermediatePlace(curr,MP)
10:  TR← CreateTransitions(curr,MP,IM)

11:  If curr.CA ∉ {Failure, Cancel, Refuse, not-understood}
12:          AR← CreateArcs(curr,MP,IM,TR)
13:  Else // MP is a terminating place
14:          AR← CreateArcs(curr, IM,TR)
15:  End If

16:  FixColor(MP,TR,AR,curr.CA)

17:  CPN.places ← CPN.places ∪ MP ∪ IM
18:  CPN.transitions ← CPN.transitions ∪ TR
19:  CPN.arcs ← CPN.arcs ∪ AR
20: End while
21: Return CPN
```

Fig. 2. IP to CPN Conversion Procedure

4 A Case Study: The Agent-Based Transportation e-Market System

To illustrate this algorithm, we use it to construct a CPN of a part of our example presented in [7] (shown as IP in figure 3). This example illustrates the interaction among three parts: Customer, Broker and IRevise, where the two first parts are Interfaces of different business systems, and the last part is an automatic service. In this protocol, the process starts when the Customer role sends a message with business information: `request(ItineraryData)`. Once the Broker receives these messages, the Web service IRevise is invoked for reviewing the customer itinerary and divide this itinerary into sub-itineraries.

We note that all the private processes are not defined by the interaction protocol because they are private aspects of the Broker. After dividing the itinerary, the Broker decides whether to send a message `propose(ItineraryPlan)` to the Customer or refuse the customer request because it cannot be satisfied. This is defined with a logical connector XOR, which represents that only one of the two alternative messages can be

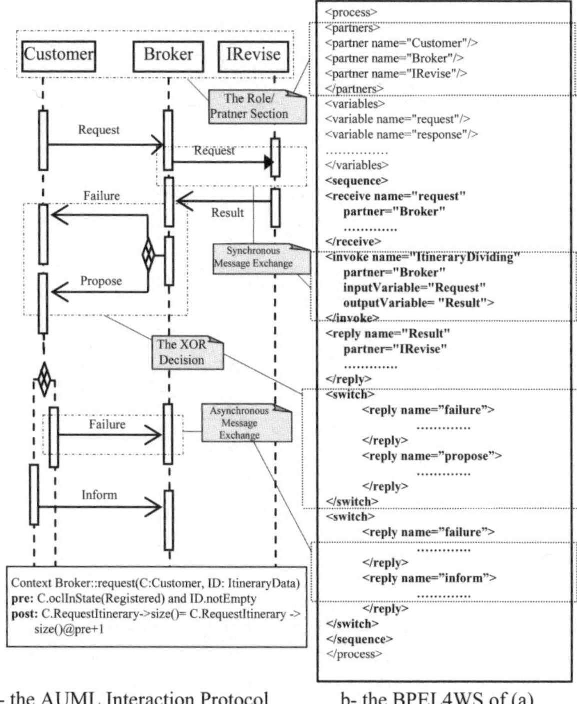

a- the AUML Interaction Protocol b- the BPEL4WS of (a)

Fig. 3. An Interaction Protocol as AUML/BPEL4WS

sent. In this case, the Customer has two interaction threads that represent the incoming messages. When the Customer receives a message `propose(ItineraryPlan)`, he can accept this itinerary plan or can declare a failure during the negotiation because consensus has not been achieved.

We now use the algorithm introduced above (fig. 2) to create a CPN for this IP. The algorithm begins with the creation of three Roles Places (RP) initially marked (one place for every role/partner in the IP: lines 3 and 4). Line 5 permits to update the CPN with the *RP* variable. In the first iteration of the main loop (line 7), the *curr* variable is set to the first message in the IP (*curr* $\leftarrow$ *<"Customer", "Broker", "request", "S">*). The algorithm creates net places, which are associated with the *curr* variable, i.e. a request message place (line 8) and two places in the Customer and respectively the Broker sub-nets (the `CreateIntermediatePlace()` function at line 9).

These three places (see the resulting CPN in Figure 4) are connected using the asynchronous message building block shown in Table 1. The MP is not a terminating place (the Customer is waiting for a response from the Broker) and is thus connected through transitions and arcs with the `CreateTransitions()` and `CreateArcs()` functions (lines 10, 11, 12). Next, the colour sets of the corresponding places are determined (colour domains of the transitions are generally defined according to the domains of the results of functions evaluation of input arcs).

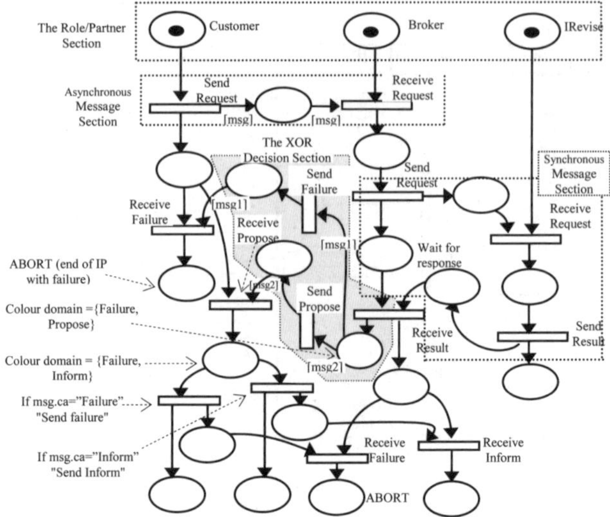

Fig. 4. The Resulting CPN of the IP presented in Fig. 3

In the second iteration, *curr* is set to *<" Broker", "Customer", "failure", "A">⊕ <" Broker", "Customer", "propose", "A">*. In this case, the Broker can send either a failure or a propose messages, and thus appropriate message places are created using the XOR-decision building block shown in Table 1. Then, two places, corresponding to the results of the messages are created. These places are connected using the XOR-decision described in Table 1. This building block involves the creation of the guard conditions on the transitions controlling the firing of the transition corresponding to the message type (which is represented as a colour in the Petri net).

In this iteration, we note that the MP place corresponding to the message "failure" is a terminating place, so no outgoing transitions or arcs are creating from this place. The loop iterates as long as M contains messages that have not been handled. Finally, the resulting CPN is returned (Figure 4).

5 Validation and Property Verification

CPN allow us to validate and evaluate the usability of a system by performing automatic and/or guided executions. These simulation techniques can also carry out performance analysis by calculating transaction throughputs, etc. Moreover, by applying other analysis techniques it is possible to verify static and dynamic properties in order to provide the complement to the simulation. Some of these properties are that:

- There are no activities in the system that cannot be realized (dead transitions). If initially dead transitions exist, then the system was bad designed.
- The IP specification exhibits the liveness property (e.g., the output CPN guarantees the existence of an initial state such that for any accessible state, at least one operation is executed).

- It is always possible to return to a previous state (home properties). For instance, to compare the results of applying different decisions from the same state. (the case of XOR and OR decision)
- The system may stop before completion (deadlock). Thus, a work might never be finished, or it might be necessary to allocate more resources to perform it.
- Certain tokens are never destroyed (conservation). Hence, resources are maintained in the system.

6 Enabling Integration Process with Multi-Agent Systems

As we already have said, the BPEL4WS process specification is considered as a language for specifying the interaction protocol of multi-agents system. In this section we briefly describe how the MAS use the verified and validated BPEL4WS specification to establish the BPI. Our suggestion consists in the addition of a specific agent between the MAS application and its IP parts conceived as Web services (see figure 5). The main advantage of this approach is the integration completeness property inherent from our BPEL4WS specification. Integration completeness means that the IP is itself published and accessed as a Web service that can participate in other application integration. Since applications integration is often viewed as a hierarchy of different local systems and services, the integration completeness property permits the agent-based integration to be included via BPEL4WS into other applications integration definitions.

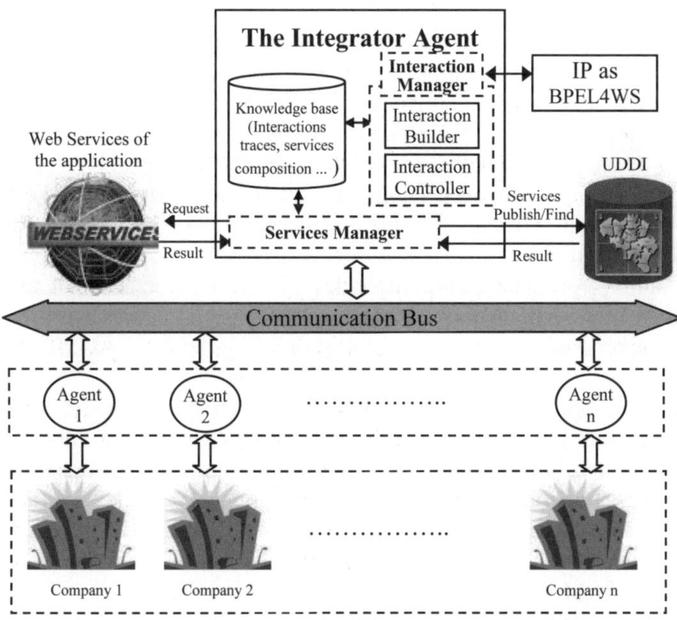

Fig. 5. Global Structure of our Architecture

As shown in Figure 5, the BPEL4WS specification is exploited thereafter with an intermediate agent called «Integrator Agent ». This integration must keep as much as possible the autonomy of architecture core based on agents. Indeed, The agents are coordinated with the Integrator agent and the exchange of messages to enact the BPI. In this architecture, the following communication pathways exist:

- agent to agent communication occurs via FIPA ACL and is facilitate by a FIPA compliant Agent Management System.
- agent to Web service communication is accomplished via SOAP messages.
- agent to BPEL4WS dataspace communication uses appropriate protocols/interfaces provided by the dataspace. The dataspace is used to store BPEL4WS process variables, which maintain the state of the IP.

The main roles of the Integrator agent are the creation, monitoring, and control of IP life cycle. It's architecture features two modules: an interaction manager and a service manager. The interaction manager contains operational knowledge (e.g., Interactions states). It also provides operations for monitoring interactions (i.e., creating and deleting instances). The service manager provides methods for receiving service requests, tracing service executions, and communicating with service requesters in accordance with IP definition (e.g., sending a notification informing the requester that deadline for cancelling an operation is passed).

7 Related Work

BPI and automation is an active research domain. The community is still debating the issues of enterprises collaboration at the business process level.

In [17], P. Buhler et al. summarize the relationship between agents and Web services with the aphorism "Adaptive Workflow Engines = Web Services + Agents": namely, Web services provide the computational resources and agents provide the coordination framework. They propose the use of the BPEL4WS language as a specification language for expressing the initial social order of the multi-agent system. P. Buhler et al. does not provide any design issues to ensure the correctness of their interaction protocols.

In [21], authors propose translating rules for the conversation of an interaction protocol given in AUML to CPN. Unfortunately, no procedures were provided that guide the conversion of an interaction protocol given in AUML to Petri net representations.

The Symphony project [18] has developed an algorithm for analyzing a composite service specification for data and control dependences and partitioning it into a set of smaller components. These components are then distributed to different locations and, when deployed, cooperatively deliver the same semantics as the original workflow. Symphony does not provide any support for failures arising from workflow mismatches since it assumes that the distributed processes will be derived from a single complete BPEL process.

Several other approaches aim to solve the integration problem by emphasizing interaction protocols. The state transition diagram (STD) has been extensively used for IP specification due to its clarity. The weakness is that it does not reflect the asynchronous character of the underlying communication [19]. Furthermore, it is not easy to represent integration of protocols. The Dooley Graph [20] is an alternative formalism for visualizing agent inter-relationships within a conversation. Object-oriented methods like UML [22] offer a way to reduce the gap between users and analyst when considering message transfers, yet they only address the dynamic behavior of individual objects and are informal.

Compared with the related work, our approach allows us to provide a clear separation of inter-enterprise collaboration management and local business process management, to make full use of existing workflow system components, to support both public processes and private business processes. Another advantage of our approach is the integration completeness property inherent from our BPEL4WS specification. It means that the IP is itself published and accessed as a Web service that can participate in other application integration. Since applications integration is often viewed as a hierarchy of different local systems and services, the integration completeness property allows agent-based integration to be incorporated via BPEL4WS into other applications integration definitions.

8 Conclusion and Future Work

In this paper, we presented a generic approach for BPI based on interaction protocols. The proposed translation rules from AUML/BPEL4WS notations to Coloured Petri nets enable the use of many verification techniques during the design phase to detect errors as early as possible.

Such translation allows to easily model complex e-business applications. We also proposed an automated procedure for converting an interaction protocol specification to a corresponding coloured Petri nets and illustrated its use through a case study.

The verified and validated interaction protocols specification is exploited afterwards with an intermediate agent called Integrator Agent to enact the integration process and to manage it efficiently in all steps of composition and monitoring.

Our primary future work direction is the exploitation of the BPEL4WS specified BPI by the Integrator agent to facilitate the creation, monitoring, and control of interaction life cycle at run-time. We will also introduce the notion of intelligence; we will try to specify all the cooperative agents of our architecture as intelligent and autonomous Web components.

References

1. Papazoglou, M.P., Kratz, B.: Web Services Technology in Support of Business Transactions. Int. journal of Service Oriented Computing 1(1), 51–63 (2007)
2. Jung, J.Y., Kang, S.H.: Business Process Choreography for B2B Collaboration. IEEE Internet Computing, 37–45 (2004)
3. Aissi, S., Malu, P., Srinivasan, K.: E-business process modeling: the next big step. IEEE Computer, 55–62 (2002)
4. Luck, M., McBurney, P., Shehory, O., Willmott, S.: The AgentLink Community: Agent Technology: Computing as Interaction - A Roadmap for Agent-Based Computing. AgentLink III (2005)

5. Benmerzoug, D., Boufaida, M., Kordon, F.: A Specification and Validation Approach for Business Process Integration Based on Web Services and Agents. In: Int. Workshop on Modeling, Simulation, Verification and Validation of Enterprises Information Systems (MSVVEIS 2007), pp. 163–168. INSTICC press (2007)
6. Benmerzoug, D., Boufaida, Z., Boufaida, M.: From the Analysis of Cooperation Within Organizational Environments to the Design of Cooperative Information Systems: An Agent-Based Approach. In: Meersman, R., et al. (eds.) OTM Workshops 2004. LNCS, pp. 496–506. Springer, Heidelberg (2004)
7. Benmerzoug, D., Boufaida, M., Boufaida, Z.: Developing Cooperative Information Agent-Based Systems with the AMCIS Methodology. In: IEEE International Conference on Advances in Intelligent Systems: Theories and Application, Luxembourg (2004)
8. Cost, R., Chen, Y., Finin, T., Labrou, Y., Peng, Y.: Using Colored Petri nets for Conversation Modeling. In: Dignum, F., Greaves, M. (eds.) Issues in Agent Communication. LNCS (LNAI), vol. 1916, pp. 178–192. Springer, Heidelberg (2000)
9. Girault, C., Valk, R.: Petri Nets for Systems Engineering - A Guide to Modeling, Verification, and Applications. Springer, Heidelberg (2003)
10. Koehler, J., Tirenni, G., Kumaran, S.: From Business Process Model to Consistent Implementation: A Case for Formal Verification Methods. In: Pro. of the Sixth International Enterprise Distributed Object Computing Conference, IEEE Computer Society, Los Alamitos (2002)
11. Peregrine B2B Integration Platform, http://www.peregrine.com
12. Thatte, S.: XLANG: Web Services for Business Process Design, Microsoft Corp., cf (2001), http://www.gotdotnet.com/team/xml_wsspecs/
13. Huget, M., Odell, J.: Representing agent interaction protocols with agent UML. In: 3rd International Joint Conference on Autonomous Agents and Multiagent Systems, pp. 1244–1245. IEEE Computer Society, Los Alamitos (2004)
14. Business Process Execution Language for Web Services Version 1.1 (2003), http://www-106.ibm.com/developerworks/
15. Sun Microsystems. Java Web Services Development Pack 1.1 (2006), http://java.sun.com/webservices/webservicespack.html/
16. Gutnik, G., Kaminka, G.A.: A Scalable Petri Net Representation of Interaction Protocols for Overhearing. In: Kudenko, D., Kazakov, D., Alonso, E. (eds.) AAMAS 2004. LNCS (LNAI), vol. 3394, pp. 1246–1247. Springer, Heidelberg (2005)
17. Buhler, P.A., Vidal, J.M.: Towards adaptive workflow enactment using multiagent systems. Int. Jour. On Information Technology and Management, 61–87 (2005)
18. Chafle, G., Chandra, S., Mann, V., Nanda, M.: Decentralized Orchestration of Composite Web Services. In: Proc. of the Alternate Track on Web Services at the 13th International World Wide Web Conference (WWW 2004), pp. 134–143 (2004)
19. Martial, F.: Coordinating Plans of Autonomous Agents. LNCS (LNAI), vol. 610. Springer, Heidelberg (1992)
20. Parunak, H.V.D.: Visualizing Agent Conversations: Using Enhanced Dooley Graphs for Agent Design and Analysis. In: Proceedings of the International Conference on Multi-Agent Systems (1996)
21. Mazouzi, H., Fallah-Seghrouchni, A.E., Haddad, S.: Open Protocol Design for Complex Interactions in Multi-Agent Systems. In: Proceedings of AAMAS 2002, pp. 517–526 (2002)
22. Booch, G., Rumbaugh, J., Jacobson, I.: The unified modeling language for object-oriented development. Document set version 1.0, Rational Software Corporation, Santa Clara (1997)
23. OMG; Object Constraint Language Specification, http://www.omg.org/cgi-bin/doc?formal/03-03-13

Competencies and Responsibilities of Enterprise Architects

A Jack-of-All-Trades?

Claudia Steghuis[1] and Erik Proper[1,2]

[1] Capgemini, Papendorpseweg 100, 3500 GN Utrecht, The Netherlands
claudia.steghuis@capgemini.com
[2] Radboud University Nijmegen, Toernooiveld 1,
6525 ED Nijmegen, The Netherlands
e.proper@acm.org

Abstract. This paper is not concerned with *enterprise architecture* as a product or as a process, but rather concerns itself with the professionals who are responsible for the creation of the products and the execution of the associated processes: *the enterprise architects.*

We will discuss the responsibilities of enterprise architects, as well as the basic competencies and personality types which an enterprise architect is expected to have in meeting these responsibilities. Since enterprise architects are likely to operate in teams we also discuss the competencies needed to effectively work in teams.

The presented results are based on existing studies into the skills of architects, surveys conducted among enterprise architects, as well as the experience of our organisations in teaching future enterprise architects.

Keywords: Enterprise architecture, competencies.

1 Introduction

The emerging instrument of *enterprise architecture* promises to provide management with insight and overview to harness the complexities involved in the evolution and development of enterprises. Where classical approaches will handle problems one by one, *enterprise architecture* aims to deal with these issues in a coherent and integral fashion. At the same time it offers a medium to achieve a shared understanding and conceptualisation among all stakeholders involved and govern the enterprise's evolution and development based on this conceptualisation. This paper focusses on the person who needs to execute these tasks; *The Enterprise Architect.* We aim to discuss the competencies, roles, and abilities needed by an enterprise architect to best conduct their tasks.

One only needs to look at one of the many job-adds to see that an enterprise architect needs to have a wide range of competencies. Consider for example:

> *Assist the Enterprise Architecture team to develop a Target EA, Transition Plan and EA governance strategies. Work with lead to develop all stages*

J.L.G. Dietz et al. (Eds.): CIAO! 2008 and EOMAS 2008, LNBIP 10, pp. 93–107, 2008.

of enterprise architecture, information engineering, system development methodologies, EA strategic planning, business process re-engineering, workflow processing, requirements analysis, prototyping, system testing, major system and database implementation. Assist in the development of an EA roadmap and strategy, current architecture assessment, architecture tools and repository evaluation and approach, development of EA governance, communication, metrics, investment management, modelling of current and target architecture views, gap analysis, and migration plan to integrate their IT efforts with mission goals.[1]

This example shows that the role of enterprise architect demands leadership qualities, a deep knowledge of IT and business domains as well as ample communication skills. Clearly not a starter's position. In randomly chosen job adds for enterprise architects, the following tasks and responsibilities are asked for:

- *Responsible for executing the architectural vision for IT systems within the organisation including those that support Internet applications, ensuring that architecture conforms to enterprise standards.*
- *Provide technical and architectural direction to the software and infrastructure team.*
- *Stay constantly attuned to emerging technologies and recommend business direction based on those technologies.*
- *Provides technical expertise to peers and associates on overall distributed enterprise architecture and design.*
- *Assist in developing and maintaining strategies that result in efficient and effective use of enterprise core services.*
- *Strong conceptual and analytical skills.*
- *Experience in creating and defining new technology concepts and solutions.*
- *Java development experience preferably in a SAP Enterprise Portal environment.*
- *Experience in development of Segment Architectures that align with and enable agency strategic goals and business requirements.*

The requirements put on an enterprise architect seem to range from very specific programming skills to broad leadership qualities as well as the ability to develop a business strategy. Tasks and responsibilities differ per job add: there is no one set of tasks and responsibilities for the role of enterprise architect.

Besides an *enterprise architect*, there are many other types of architects, such as business architects, information architects, process architects, IT architects, software architects, application architects, etcetera. The difference between these types of architects and the enterprise architect is that the enterprise architect covers the breadth of business and IT, while domain architects focus on one aspect of the enterprise (business, IT, information) and solution architects on one small part of the implementation of the architecture (applications, software, business processes).

[1] From: `http://hotjobs.yahoo.com/jobseeker/jobsearch/job_detail.html?job_id=JVVWL53A4E1` (11-01-2008).

Some initial work has already been done regarding the abilities and competencies that should be met by enterprise architects. For instance, organisations such as TOGAF [1] and the Netherlands Architecture Forum [2] have created frameworks of competencies for architects. Some organizations have created their own competencies frameworks [3, 4], or have even introduced their own certification programs (for example: IBM, HP, Capgemini, Federal Enterprise Architecture Certification Institute, and TOGAF).

Standard guidelines regarding the competencies of an enterprise architect still lack. Responsibilities differ per company/assignment and research showed that architects themselves expect to have to have a variety of competencies [5]. Using pre-existing frameworks for competencies and abilities [6, 7, 8, 1, 2, 5, 3, 4] as a starting point, this paper provides a competency framework for enterprise architects which is geared towards the responsibilities of enterprise architects.

The goal of our study is to collect information about the profile of the enterprise architect, to be able to improve the composition of enterprise architecture teams and to improve education programmes for future enterprise architects. Therefore, we compared earlier research about competence frameworks with literature about enterprise architecture roles and Belbin team roles. We have synthesized a framework in which we relate: the responsibilities of enterprise architects, relevant competencies, typical roles of an architect, as well as their roles in teams, by means of a number of mappings. These actual mappings are the result of studying earlier work on competencies of enterprise architects [2,4], as well as a survey conducted among (certified) enterprise architects from Capgemini. With our research we aim to answer the following questions:

1. What competencies do enterprise architects need in meeting their responsibilities?
2. What roles/attitudes should enterprise architects cover and what competencies are needed for those roles?
3. What team roles should be fulfilled by enterprise architects?

This paper is structured as follows. In section 2 we discuss the basic competencies which an enterprise architect is expected to have, while section 3 summarises the responsibilities of enterprise architects and relates them to the competencies. In section 4 we then continue by discussing the personality types needed to meet the responsibilities enterprise architects. Since enterprise architects are likely to operate in teams, section 5 considers competencies related to working in teams.

2 Relevant Competencies

In this section we look at the competencies that are relevant to the work of enterprise architects. As we will see in the next section, not all of these competencies are relevant to each of the roles played by architects.

According to a survey among enterprise architects, one has to be a jack-of-all-trades to be a good enterprise architect [5]. Even more, job adds for enterprise architects typically claim at least five years of experience, profound domain expertise, specific knowledge about networks, applications, operating systems, etc, communication skills and proven success in implementation. Providing a complete list of competencies of the architect is therefore also hardly possible. We will limit ourselves by introducing the essential competencies on the different fields which are needed. In doing so, we distinguish two kinds of competencies:

Professional competencies – Competencies dealing with knowledge, attitude and skills necessary to a successful performance in a specific function or role [9].

Personal competencies – Competencies that can be used in several functions or roles (i.e. communication skills) and personality characteristics.

2.1 Professional Competencies

The professional competencies comprise the knowledge, attitude and skills to perform successfully in a specific function [2]. The enterprise architect should be able to understand and have knowledge of all four areas (business, information, information systems and infrastructure), while he needs to be an expert in at least one area [3]. TOGAF divides the professional competencies in their Architecture skills framework in business skills and methods, enterprise architecture skills, programme or project management skills, IT general knowledge skills, technical IT skills, and legal environment [1].

When looking at the competence model of a standardisation effort such as TOGAF [1] as well as the competence model of an architecture society such as the NAF [2], one can conclude that architects need to have knowledge about the different domains they act in. In addition, knowledge about architecture principles, architecture frameworks and governance is most important, while keeping informed about new developments is also necessary.

2.2 Personal Competencies

For the personal competencies we do not distinguish between different types of architects. Even more, it seems those competencies are quite close to adjacent professions such as strategists, process developers and system developers. The personal competencies can be divided in intermediary competencies, values, norms and ethics and personality characteristics [2]. This last group contains natural abilities of a person and these are therefore hard to be learned. One of these is persuasiveness, which is recognized by [7] as an important characteristic of an architect. Others are independence, persistence, initiative, etcetera [2]. Values, norms and ethics differ per person and organization. Intermediary competencies are the ones mostly mentioned in literature and job adds. A short comparison between four sources [1,2,10,11] showed the following top five intermediary competencies for the architect (using the naming conventions of [2]):

- Analytical skills.
- Communication skills.
- Negotiation.

- Abstraction capacity.
- Sensitivity and empathy.

Besides those, creativity and leadership appear to be essential for the enterprise architect, especially because (s)he needs to cover the whole spectrum of business and ICT and often operates in a leadership role in close collaboration with other architects. Based upon [2] and extended with some competencies concerning change management from [8] and communication [3], we identify the following personal competencies:

Abstraction capacity – The ability to learn in new situations and to adapt acquired knowledge and facts, rules, principles to new domains.

Accurateness – Working neatly and precise.

Analytical skills – The ability to identify a concept or problem, to dissect or isolate its components, to organise information for decision making, to establish criteria for evaluation, and to draw appropriate conclusions.

Authenticity – Being true to one's own personality, spirit, or character.

Consulting – Being able to give recommendations on a certain case.

Creativity – To be able to generate creative ideas and solutions, invent new ways of doing business, and be open to new information.

Decisiveness – To be able to take decisions after having enough or complete information and act towards these decisions.

Dedication – Driven to accomplish their goals.

Didactical skills – The ability to transfer complex knowledge to other people.

Diplomacy – Ability to communicate about sensitive issues without arousing hostility.

Facilitation skills – Be able to facilitate workshops.

Flexibility – Ability to deal with changed conditions, assumptions, environment, etc.

Independency – To be able to act without being influenced by others.

Initiative – Readiness to act on opportunities.

Integrity – Moral soundness.

Leadership – Inspiring and guiding groups and people.

Listening – Listen actively to understand information or directions and be able to provide relevant feedback.

Loyalty – Faithful to the key stakeholders.

Negotiation – To be able to maintain a position in conversation with others and improve this position.

Openness – Open to alternative directions, solutions and opinions.

Opinion forming – Being able to make a judgement about a certain case.

Organisational awareness – To understand the inner working of the organisation; to estimate the value of the own influence and consequences of decisions or activities.

Persistence – Being determined to do or achieve something.

Persuasiveness – To be able to convince others of a certain opinion.

Plan and organize – Making objectives and take actions to reach these objectives in an effective way.

Result driven – To be able to realise objectives and results.
Self-confident – Confident about (and familiar with) their own (in)abilities.
Self-development – Reflect on your performance and goals, identify learning needs and development options, and develop knowledge and skills.
Sensitivity and empathy – Sensing others' feelings and perspective, and taking an active interest in their concerns.
Stability – Has a stable character and mood.
Teamwork – Working with others towards shared goals and creating group synergy in pursuing these goals.
Verbal communication skills – Use appropriate technical or business vocabulary to be able to express thoughts and feelings in a concise way and to respond adequately to others.
Visualisation skills – Be able to visualize architecture results.
Working systematically – Be able to execute the work in a prescribed way.
Written communication skills – Write clear and accurate reports, letters and documents.

3 Responsibilities of an Enterprise Architect

According to [11] an enterprise architect's job can involve governance committees, architecture review boards, technology life cycles, portfolio management, architecture strategy and strategic project support. Bredemeyer [7] shows that enterprise architecture has broadened its scope from just an IT issue to the enterprise wide IT architecture and business architecture, with as goal to increase enterprise agility and alignment with business strategy.

In [12] a more elaborate discussion of the process of architecting and the responsibilities architects have in this process is provided. This discussion is based on a survey involving several sources, such as: [13,14,15,16,17,18,19,20,1, 7,21,22,23]. In this paper we will only provide a summary of the responsibilities of an enterprise architect:

Create: The creation of an enterprise architecture.

- Understand purpose and context of the enterprise architecture.
- Determine which deliverables are required for the creation of a specific enterprise architecture.
- Monitor the enterprise's context and the stakeholders involved in the enterprise's development.
- Create shared conceptualisation among the stakeholders involved in the enterprise's development.
- Design the processes involved in creating the enterprise architecture.
- Determine impacts of alternative enterprise architectures.
- Communicate results of the creation process.

Apply: The application of an enterprise architecture.

- Inform stakeholders about the selected enterprise architecture and its motivations.
- Support decision-making based-on the enterprise architecture.
- Ensure compliance of development of the enterprise to the architecture.
- Ensure the enterprise architecture results are available.
- (Re)-communicate the architecture and its impact to relevant stakeholders.

Maintain: The maintenance of an enterprise architecture.

- Monitor the enterprise's context and the stakeholders involved in the enterprise's development.
- Assess drivers for change inside/outside the enterprise.
- Update and (re-)communicate the enterprise architecture.

Organise: The organisation of the processes involved in enterprise architecting.

- Organise the enterprise architecture team.
- Select frameworks, tools and tricks.
- Communicate about enterprise architecture as a means.
- Embed enterprise architecting in the enterprise's governance.
- Monitor maturity of the enterprise architecting process.
- Manage quality of the enterprise architecture; both product and process.
- Establish leadership.
- Innovate the architecture processes.

In meeting these responsibilities, the enterprise architect needs certain personal competencies. Table 1 provides a mapping from the responsibilities to the competencies discerned in the previous section based upon a survey among (certified) enterprise architects within Capgemini. We have not mapped the professional competencies to the responsibilities, this needs further research.

4 Personality Types

Strano et al. [24] report on a survey conducted among enterprise architects of the federal government of the United States of America, and concluded that an enterprise architect can have the roles of a *change agent*, *communicator*, *leader*, *manager*, and *modeller*. In [24] these roles are defined as:

Change agent – "*As a change agent, the enterprise architect supports enterprise leaders in establishing and promoting the best strategy to accomplish business goals and objectives.*"

Communicator – "*As a communicator, he assists managers, analysts, systems architects and engineers in understanding the details of the strategy sufficiently well to make decisions and execute the plan that leads to realization of the shared vision.*"

Table 1. Competencies mapped upon responsibilities

Competences \ Architecture Process	*Create*	Understand purpose and context	Determine deliverables	Monitor context and stakeholders	Create shared conceptualisation	Design creation process	Determine impacts	Communicate	*Apply*	Inform	Support decision-making	Ensure compliance	Make results available	(Re-)communicate	*Maintain*	Monitor context & stakeholders	Assess drivers for change	Update & Communicate	*Organise*	Organise team	Select framework, tools & tricks	Communicate about EA	Embed EA in Governance	Monitor maturity	Manage quality	Establish leadership	Innovate
Abstraction Capacity		x			x	x	x				x	x									x						
Accurateness		x	x		x	x	x			x	x	x	x				x	x							x		
Analytical Skills		x	x	x	x		x				x	x				x	x	x			x		x				x
Authenticity					x	x		x		x	x			x						x		x				x	
Consulting		x	x	x	x	x	x	x		x	x	x		x		x		x		x	x	x	x			x	
Creativity			x		x	x		x			x			x							x	x	x				x
Decisiveness		x	x	x	x	x					x	x					x			x	x				x	x	
Dedication			x	x	x								x	x		x	x	x						x	x		
Didactical Skills			x	x				x		x			x	x								x		x		x	
Diplomacy		x		x	x			x		x	x	x		x								x	x			x	
Facilitation skills		x			x			x			x	x	x	x		x		x		x		x	x				
Flexibility				x	x	x														x						x	
Independency		x		x	x					x	x	x				x	x							x	x		x
Initiative		x	x			x		x		x	x	x	x	x		x	x	x		x	x	x	x	x	x	x	x
Integrity		x	x	x	x			x			x	x											x	x	x	x	
Leadership		x	x	x	x	x					x	x						x		x	x	x	x		x	x	
Listening		x		x	x			x			x	x		x		x						x				x	
Loyalty				x								x								x						x	
Negotiation				x	x			x			x	x								x			x		x		
Openness		x			x			x		x	x	x	x			x				x		x					
Opinion Forming		x	x	x				x				x				x	x				x			x	x		x
Organisational Awareness		x		x		x	x	x		x	x	x		x		x	x			x		x	x	x	x	x	
Persistence			x	x	x			x				x					x					x	x	x	x		
Persuasiveness			x		x	x		x				x								x		x	x			x	
Plan And Organize			x			x							x							x			x	x	x	x	x
Result Driven		x	x		x	x					x		x	x				x		x	x		x		x		
Self Development		x																						x			x
Self-confidence					x		x	x		x	x	x	x	x						x	x	x				x	x
Sensitivity And Empathy		x		x	x			x		x	x	x		x						x		x	x		x	x	
Stability			x		x	x	x	x			x	x								x		x	x	x	x	x	
Teamwork		x	x	x	x	x					x									x		x		x		x	x
Verbal Communication		x	x	x	x			x		x	x	x	x	x		x		x				x	x		x	x	x
Visualisation skills					x	x		x			x		x	x				x									x
Working Systematically		x	x	x		x	x				x	x	x			x	x	x			x				x		
Written Communication			x	x	x		x	x		x	x		x	x				x				x	x		x	x	x

Leader – *"As a leader, the enterprise architect participates in creating a shared vision, motivating members of the enterprise to aspire to achieving the vision, and providing clear direction regarding what is required to execute a strategy to accomplish goals and objectives that result in performance improvements."*

Manager – *"As a manager, he organizes the architecture team and ensures that adequate resources are secured to perform the architecture process."*

Modeller – *"As a modeller, the enterprise architect provides a representation of the relationships of enterprise components with sufficient detail and in the format needed to enable making necessary decisions to execute the strategic plan."*

As an alternative to these roles, [7] suggests four competency areas: credible expert, strategist, politician and leadership. In this paper we adapt the roles of [24] since they are based on a documented empirical study.

In [8] five stereotypical styles of thinking about change are identified. Each style is typed by its own-colour:

Blueprint-thinking – Focuses on the formulation of unambiguous objectives, development of a plan of action, monitoring and adjusting the change process accordingly.

Yellowprint-thinking – Focuses on bringing interests together, stimulating stakeholders to formulate opinions, creating win-win situations and forming coalitions.

Redprint-thinking – Focuses on stimulation of people, and implementing sophisticated HRM-instruments.

Greenprint-thinking – Focuses on ensuring that people are aware of new perspectives and personal shortcomings, while motivating them to see, learn, do new things, and create suitable shared learning experiences.

Whiteprint-thinking – Focuses on the natural flow of people's processes, interests and energies, and is concerned with the removal of blockades.

Each of these "colours" of thinking about change has their own merits. Depending on the organizational culture and architectural maturity in which an enterprise architect needs to operate, a different prevailing style will be needed.

The five roles from [24] can be mapped upon the competencies mentioned in section 2 In most of these roles, communication, negotiation and sensitivity and empathy play a large role. Analytical skills and abstraction capacity are definitely needed for the modeller, but are also important to fulfil such a multidimensional role as enterprise architect. Using the competencies of enterprise architects as discussed in the previous section, these roles can be made more specific as shown in Table 2. Note that we have treated the roles as "extremes" or "caricatures" when mapping the competencies. For example, to be a leader, an architect will also need some abstraction capacity. Nevertheless, the ability to abstract is really the core of their role as modeller. Conversely, when modelling, an architect also needs to be able to listen, which is a key trait for the communicator role.

In Table 2, the *change agent* role has been refined to include the colours of thinking about change discussed in [8]. In this table, we can see that the first four

Table 2. Mapping competencies to roles and change colours

Competencies \ Roles	Communicator	Leader	Manager	Modeller	Change Agent: Yellow Change Agent	Blue Change Agent	Red Change Agent	Green Change Agent	White Change Agent
Abstraction Capacity				x					
Accurateness			x	x		x			
Analytical Skills				x		x			
Authenticity		x							
Consulting	x	x							
Creativity		x		x				x	
Decisiveness		x	x			x	x		
Dedication		x		x		x	x		
Didactical Skills	x	x	x					x	
Diplomacy			x		x				
Facilitation skills	x		x						
Flexibility		x			x		x	x	x
Independency		x	x		x	x			x
Initiative		x	x						
Integrity		x	x				x		
Leadership		x	x		x				
Listening	x			x				x	
Loyalty									
Negotiation	x		x		x				
Openness									
Opinion Forming		x							x
Organisational Awareness			x		x	x	x	x	x
Persistence		x	x		x				
Persuasiveness	x	x			x		x		
Plan And Organize			x			x			
Result Driven			x						
Self Development		x						x	
Self-confidence		x			x			x	x
Sensitivity And Empathy		x	x				x	x	x
Stability		x	x		x				
Teamwork	x	x	x				x		
Verbal Communication	x	x	x	x	x	x	x	x	x
Visualisation skills	x			x					
Working Systematically			x	x		x			
Written Communication	x	x	x	x		x			

roles have many competencies in common, while the modeller is a completely different role.

Combining Table 1 with Table 2 results in Table 3. When examining this latter table, it is most striking to see that responsibilities and roles are not aligned to each other. Some responsibilities are attached to no role at all, while others are a combination of all roles. This really calls for future research. What seems to be the case is:

- No justice is done to the responsibilities involved in the maintanance of architectures. At the moment, only the modeller and blue change agent role are important for these.
- The communicator role seems less necessary than expected.

Table 3. Relating process and responsibilities to roles

Architecture process (Ch. 5) \ Role	Communicator	Leader	Manager	Modeller	Change Agent: Yellow	Change Agent: Blue	Change Agent: Red	Change Agent: Green	Change Agent: White
Create									
Understand purpose and context			x	x		x	x		x
Determine deliverables	x	x	x	x		x	x		
Monitor context and stakeholders	x	x	x	x	x	x	x	x	x
Create shared conceptualisation	x	x	x	x	x	x	x	x	x
Design creation process									
Determine impacts									
Communicate	x	x	x		x			x	x
Apply									
Inform									x
Support decision-making	x	x	x	x	x	x	x	x	x
Ensure compliance	x	x	x	x	x	x	x		x
Make results available				x		x			
(Re-)communicate	x			x				x	
Maintain									
Monitor context & stakeholders						x			
Assess drivers for change						x			
Update & Communicate				x		x			
Organise									
Organise team					x		x		
Select framework, tools & tricks									
Communicate about EA	x	x	x		x			x	
Embed EA in Governance	x		x		x				
Monitor maturity									
Manage quality			x		x	x	x		x
Establish leadership	x	x	x		x		x	x	x
Innovate									

5 Enterprise Architecture Teams

Since enterprise architects are likely to operate in teams, it is not necessary to find a single person who fulfils all competencies. To combine a team of architects it is not only necessary to find a good coverage of the competencies defined in section 2, but also to ensure the group of selected architects indeed operates as a team. It is therefore also relevant to consider models for the abilities of people to work in teams.

In [6] a number of roles of members in teams are identified:

Implementer – Well-organised and predictable. Takes basic ideas and makes them work in practice. Can be slow.

Shaper – Lots of energy and action, challenging others to move forwards. Can be insensitive.

Completer/Finisher – Reliably sees things through to the end, ironing out the wrinkles and ensuring everything works well. Can worry too much and not trust others.

Table 4. Belbin roles and the architecture process

Architecture process \ Belbin-role	Completer	Coordinator	Implementer	Monitor	Plant	Resource investigator	Shaper	Teamworker
Create								
Understand purpose and context				x				
Determine deliverables	x					x	x	
Monitor context and stakeholders	x	x	x				x	
Create shared conceptualisation	x	x		x	x	x	x	x
Design creation process					x			
Determine impacts	x		x	x				
Communicate						x		x
Apply								
Inform								
Support decision-making	x	x	x	x	x	x		x
Ensure compliance	x	x	x	x				x
Make results available	x		x			x		
(Re-)communicate						x		x
Maintain								
Monitor context & stakeholders			x			x		
Assess drivers for change	x		x					
Update & Communicate	x		x					
Organise								
Organise team		x						
Select framework, tools & tricks								
Communicate about EA		x				x		x
Embed EA in Governance				x	x	x	x	
Monitor maturity	x		x					
Manage quality	x		x					
Establish leadership		x						x
Innovate								

Plant – Solves difficult problems with original and creative ideas. Can be poor communicator and may ignore the details.

Monitor/Evaluator – Sees the big picture. Thinks carefully and accurately about things. May lack energy or ability to inspire others.

Specialist – Has expert knowledge/skills in key areas and will solve many problems here. Can be disinterested in all other areas.

Coordinator – Respected leader who helps everyone focus on their task. Can be seen as excessively controlling.

Team worker – Cares for individuals and the team. Good listener and works to resolve social problems. Can have problems making difficult decisions.

Resource/investigator – Explores new ideas and possibilities with energy and with others. Good networker. Can be too optimistic and lose energy after the initial flush.

Within a team of enterprise architects there should be a balance between each of these roles. When considering the responsibilities identified in section 3, one can identify shifts in the priority that should be given to each of the involvement roles. We have made an attempt to achieve a mapping between the team involvement roles and the responsibilities of an enterprise architect (team), by comparing the competencies attached to a team role with the competencies from Table 1. In creating the table, all role/responsibility combinations were selected were the team role had at least 60% of their underlying competencies in common with the competencies required by the responsibility. The result of this are shown in Table 4.

The specialist is left out of scope for the comparison, because this is the person who is needed for expert roles, and less for his personal competencies. While all roles are assigned to at least one responsibility, there are many tasks who are assigned to more than one role. Therefore, there seems to be no direct link between the roles and the responsibilities. An enterprise architect seems to be able to fulfil multiple roles for executing one responsibility. It is also striking to see that not all responsibilities are mapped to these roles. The 'Inform' responsibility somehow is not mapped to Belbin-roles.

6 Conclusion

In this paper we discussed the basic competencies which an enterprise architect is expected to have, and tied these to the personality types needed to meet the responsibilities of enterprise architects. Though this match provides insight into the responsibilities, roles and competencies of architects, further research is needed. The alignment between roles and responsibilities was not what we had expected. Some responsibilities are attached to no role at all, while others are a combination of all roles. Since enterprise architects are likely to operate in teams we also discussed the competencies needed to effectively work in teams. Also in this case, not all competencies and responsibilities were mapped.

Using the presented framework as a starting point, we aim to further investigate (mainly using surveys among enterprise architects) the responsibilities, competencies, personality types and team roles relevant to enterprise architects, as well as the mapping between these. The results of these surveys will then be used to improve our training programs for enterprise architects. In future surveys we aim to involve enterprise architects in general, as also done in the initial studies of the Netherlands Architecture Forum [2], and not only Capgemini's architects.

References

1. TOGAF: TOGAF – The Open Group Architectural Framework (2005), `http://www.togaf.org`
2. Steghuis, C., Voermans, K., Wieringa, R.: Competencies of the ICT architect. Technical report, Netherlands Architecture Forum (2005)

3. Capgemini: Architecture curriculum. Technical report, Capgemini (2007), http://academy.capgemini.com
4. Wagter, R., Witte, D., Proper, H.: The GEA architecture function: A strategic specialism. White Paper GEA-7, Ordina, Utrecht, The Netherlands, EU (2007) (in Dutch)
5. Voermans, K., Steghuis, C., Wieringa, R.: Architect roles and competencies – A questionnaire conducted during the Dutch Architectural Conference 2004. Technical report, Netherlands Architecture Forum (2005) (in Dutch)
6. Belbin, R.: Team Roles at Work. Butterworth Heinemann (1993) ISBN-10: 0750626755
7. Bredemeyer, D., Malan, R.: What It Takes to Be a Great Enterprise Architect. Enterprise Architecture - Cutter Consortium 7 (2004)
8. de Caluwé, L., Vermaak, H.: Learning to Change: A Guide for Organization Change Agents. Sage publications, London (2003)
9. Bergenhenegouwen, G., Mooijman, E., Tillema, H.: Strategic education and learning in organisations, 2nd edn. Kluwer, Deventer (1999) (in Dutch)
10. Bean, S.: The elusive enterprise architect. IT Adviser (2006), http://www.nccmembership.co.uk
11. Walker, M.: A day in the life of an enterprise architect. Technical report, Microsoft corporation (2007), http://msdn2.microsoft.com/en-us/library/bb945098.aspx
12. Op 't Land, M., Proper, H., Waage, M., Cloo, J., Steghuis, C.: Enterprise Architecture – Creating Value by Informed Governance (forthcoming, 2008)
13. Groote, G., Hugenholtz-Sasse, C., Slikker, P.: Projecten leiden: Methoden en technieken voor projectmatig werken, Het Spectrum, Utrecht, The Netherlands (1995) (in Dutch) ISBN-10: 9027497605
14. Mintzberg, H., Ahlstrand, B., Lampel, J.: Strategy safari – A guided tour through the wilds of strategic management. The Free Press, New York (1998)
15. Humphrey, W.: Managing the Software Process. The SEI Series in Software Engineering. Addison-Wesley Professional, Massachusetts (1989)
16. Sanden, W.v.d., Sturm, B.: Informatie–architectuur – de infrastructurele benadering. Panfox, Rosmalen, The Netherlands, EU (1997) (in Dutch) ISBN-10: 9080127027
17. Sitter, L.d.: Synergetisch produceren; Human Resources Mobilisation in de produktie: een inleiding in structuurbouw. Van Gorcum, Assen, The Netherlands, EU (1998) (in Dutch) ISBN-13: 9789023233657
18. Amelsvoort, P.v.: De moderne sociotechnische benadering – Een overzicht van de socio-technische theorie. ST-Groep, Vlijmen, The Netherlands, EU (1999) (in Dutch) ISBN-10: 9080138568
19. Grembergen, W.v., Saull, R.: Aligning business and information technology through the balanced scorecard at a major canadian financial group: its status measured with an it bsc maturity model (2001), http://www.hicss.hawaii.edu/HICSS_34/PDFs/OSKBE03.pdf
20. Pyzdek, T.: The six sigma handbook: The complete guide for greenbelts, blackbelts, and managers at all levels, revised and expanded edition (2003) ISBN-13: 9780071410151
21. Lankhorst, M., et al.: Enterprise Architecture at Work: Modelling, Communication and Analysis. Springer, Berlin (2005)

22. Wagter, R., Berg, M.v.d., Luijpers, J., Steenbergen, M.v.: Dynamic Enterprise Architecture: How to Make It Work. Wiley, New York (2005)
23. Dietz, J.: Enterprise Ontology – Theory and Methodology. Springer, Berlin, Germany (2006)
24. Strano, C., Rehmani, Q.: The role of the enterprise architect. Information Systems and E-Business Management 5, 379–396 (2007)

Interoperability Strategies for Business Agility

Mats-Åke Hugoson, Thanos Magoulas, and Kalevi Pessi

IT University of Göteborg
P.O.B. 8718, SE-412 96 Göteborg, Sweden
Jönköping International Business School
P.O.B. 1026, SE-551 11 Jönköping, Sweden
Mats-Ake.Hugoson@ihh.hj.se, thanos@ituniv.se, pessi@ituniv.se

Abstract. In times of increasing uncertainty and turbulence in the business environment, the concept of agility has become a new guiding principle for the change and development of enterprises. Agile business requires agile information systems and this have consequences on how the systems should interoperate. This paper describes and analyzes the impact of information systems interoperability strategies on business agility. Three interoperability strategies are identified; unification, intersection and interlinking. Cases from Swedish Health Care are used to demonstrate the application of the strategies. The conclusion is that the choice of interoperability strategy has significant impact on business agility and should therefore be analyzed and evaluated carefully. If the wrong strategy is chosen, there is a considerable risk for misalignment and expensive consequences. The challenge is to create architectural solutions for interoperability that are in harmony with the demands of the business, both in a short term and a long term perspective.

Keywords: Business Agility, Enterprise Architecture, Information Systems Architecture, Interoperability Strategy, Alignment.

1 Introduction

During the last few decades, new perspectives have evolved where organizations are viewed as dynamic and complex information environments. While many entrenched business and other organizations experience varying degrees of crises, we can note a rapid growth of new, flexible and dynamic organizations that have found new ways to compete and cooperate. New concepts, such as agile organizations, virtual organizations and imaginary organizations, and so forth have been minted to describe these new types of organizations that compete in an increasingly dynamic and complex environment.

Likewise, the information environment of business for which we create information systems and in which we use information technology (IT) is characterized by greater dynamics and complexity. Contemporary organizational boundaries are no longer clearly easily defined or delineated. This change has resulted in the organizations "external" and "internal" environments becoming more closely intertwined, increasing the complexity and dynamics. Today's information technology has

J.L.G. Dietz et al. (Eds.): CIAO! 2008 and EOMAS 2008, LNBIP 10, pp. 108–121, 2008.

opened the way to the creation of many different types of influences on organizations as well as cooperation between different organizations. One consequence of this is that the role of information technology has developed from being a means to rationalize and automate to being a means to create agile organizational forms.

Agility is probably one of the most important characteristics of businesses that have to cope with increasingly competitive environment. Agility gives enterprises the ability to sense and respond rapidly to unpredictable events and to take advantage of changes as opportunities. This means that the supporting information systems and their interoperability must be equally agile in order to achieve a close alignment between business and information systems.

Interoperability and the degree of integration between different information systems have become a major issue in IT development and decisions on heavy investments must be taken in this area. It is important that the chosen interoperability strategy is compatible with the various forms of business agility the organization demands. There is, however, sometimes a lack of understanding of the impact from alternative interoperability strategies, which increase the risk for expensive failures and undesired consequences on agility.

This paper describes and analyzes the impact of different interoperability strategies on business agility. Three interoperability strategies are identified and discussed: unification, intersection and interlinking. These strategies are analyzed further and cases from Swedish Health Care are used to illustrate these strategies. Finally consequences of the strategies on business agility and on business and information systems alignment are discussed.

2 Business Agility

The concept of agility has been discussed since the early 1990's in the area of manufacturing [1] and was introduced into IS research some year ago. In the year 2005, the International Federation of Information processing (IFIP) had a conference on "Business Agility and Information Technology Diffusion" [2]. At this conference the concept of agility was covered ranging from software development to business innovation.

Several definitions of agility may be found in the growing number of articles about agile business and information systems. Christoffer and Towill define agility as "a business-wide capability that embraces organizational structures, information systems, logistical processes and in particular, mindsets" [3]. Yusuf et al. [4] gives several examples of definitions in the area of manufacturing. They summarize the main points of the definitions of various authors as follows:

- High quality and highly customized products.
- Products and services with high information and value-adding content
- Mobilization of core competencies
- Responsiveness to social and environmental issues
- Synthesis of diverse technologies
- Intra-enterprise and inter-enterprise integration

Alberts and Hayes [5] discuss agility in general and claim that the key dimensions of agility are the following six attributes:

- Robustness: the ability to maintain effectiveness across a range of tasks, situations, and conditions.
- Resilience: the ability to recover from or adjust to misfortune, damage, or a destabilizing perturbation in the environment.
- Responsiveness: the ability to react to a change in the environment in a timely manner.
- Flexibility: the ability to employ multiple ways to succeed and the capacity to move seamless between them.
- Innovation: the ability to do new things and the ability to do old things in new ways; and
- Adaption: the ability to change work processes and the ability to change the organisation.

Many researchers consider responsiveness and adaptability as major attributes or key dimensions of agility. For example Ramasesh et al defines agility as "the capability of a manufacturing system to provide an effective response to unanticipated changes" [6]. According to Kidd [7] agility is about "…to adapt and respond quickly to changing customer requirements". Others consider flexibility as a key characteristic of agility. For example Christoffer and Towill [3] "A key characteristics of an agile organization is flexibility". One common denominator of agility is the capability to respond to changes and the ability to change in a purposeful way rather than in an ad hoc manner (purposeful changeability) [8].

3 Enterprise Architecture and Alignment

Since the 1970's, organizations are spending more and more money building new information systems. The fast growing number of systems and in many cases the ad hoc manner in which the systems were integrated have exponentially increased the cost and complexity of information systems. At the same time organizations were finding it more and more difficult to keep these information systems in alignment with business need. Furthermore, the role of information systems has changed during this time, from automation of routine administrative tasks to a strategic and competitive weapon. In light of this development, a new field of research and practice was born that soon came to be known as Enterprise Architecture. One of the pioneers introducing the concept of architecture was John Zachman [9][10]. His Enterprise Architecture Framework is probably one of the most referred both by practitioners and in researchers. A number of Enterprise Architecture Framework has evolved since Zachman first introduced his framework. For example the Open Group Architectural Framework (TOGAF) [11] and the Federal Enterprise Architecture Framework (FEAF) [12] to mention two well-known Frameworks.

Enterprise Architecture is usually divided into different categories or domains or architecture types. For example Aerts et al. identify three domains in which architecture matters [13]:

1. The business architecture defines the business system in its environment of suppliers and customers. The business system consists of humans and resources, business processes, and rules. The business architecture is derived from the business vision, goals and strategies.
2. The application architecture (or information systems architecture) details the information systems components of business and their interaction.
3. ICT platform architecture (or IT architecture) is the architecture of the generic resource layer, which describes the computers, networks, peripherals, operating systems, database management systems, UI frameworks, system service, middleware etc. that will be used as a platform for the construction of the system for the enterprise.

Some Enterprise Architecture Frameworks distinguish between Information Architecture and Application Architecture and present four architecture types: Business Architecture, Application (or Systems) Architecture, Data (or Information) Architecture and Technical (or IT) Architecture [11][13][14].

The developments in the various domains influence each other and the increasing need of business for agility to cope with changes may be provided by architectures supporting reflectivity [13]. Alignment practices must take into consideration the relation between the various architectures. In this paper we will focus on business architecture and information systems architecture and the alignment between these two.

4 Strategies for Interoperability

One of the major issues in designing and developing enterprise architecture is which degree of interoperability should there be between the various business units of the enterprise and how this should be reflected in the integration of their information systems. Interoperability is usually defined as "the ability of two or more systems or components to exchange and use information" [15]. TOGAF [11] defines interoperability as: (1) the ability of two or more systems or components to exchange and use shared information and (2) the ability of systems to provide and receive services from other systems and to use the services so interchanged to enable them to operate effectively together.

Much work has been done in defining the concept of interoperability from an Information Systems perspective. Clark and Jones [16] argue that understanding levels of organizational interoperability is also important. They propose a model of organizational interoperability consisting of five levels of organizational maturity:

Level 0 – Independent – This level describes the interaction between independent organizations. These are organizations that normally don't have any interaction and do not have common goals or purpose, but they may be required to interoperate in some scenario that has no precedent.

Level 1 – Ad hoc – At this level of interoperability only very limited organizational frameworks are in place which could support ad hoc arrangements.

Level 2 – Collaborative – The collaborative organizational interoperability level is where recognized frameworks are in place to support interoperability and shared goals are recognized and roles and responsibilities are allocated as part of on-going responsibilities however the organizations are still distinct.

Level 3 – Integrated – The integrated level of organizational interoperability is one where there are shared value systems and shared goals, a common understanding and a preparedness to interoperate. The frameworks are in place and practiced however there are still residual attachments to a home organization.

Level 4 – Unified - A unified organization is one in which the organizational goals, value systems, command structure/style, and knowledge bases are shared across the system.

Although Clark and Jones maturity model concerns inter-organizational interoperability, similar levels could also be used in understanding intra-organizational interoperability. For example Alter [17] identifies five levels of business process interoperability: common culture, common standards, information sharing, coordination and collaboration. In this paper we will mainly focus on the levels 2-4 (Collaborative, Integrated and Unified) interoperability according to Clark and Jones maturity model. When it comes to integration of information systems in order to achieve interoperability, levels 2-4 have their counterparts in the three interoperability strategies: Unification, Intersection and Interlinking [18][19]. The three strategies will be described in more detail below. Finally, we want to emphasis that both ontology and semantics are important issues of interoperability, however not explicitly treated in this paper. The focus of this paper is a business driven delineation of interoperability.

4.1 Unification Strategy

Unification is defined as the process of producing a common structure for two or more information systems [18]. A unification strategy creates a unified information space. One kind of unification means that two or more integration entities are merged into one entity (one common system principle). Another kind of unification is standardization of two or more systems with regard to their inner structure, functions and content. In this case, the systems don't merge into one physical system, rather there are many systems which are replicas of each other (replication principle). The latter situation covers physically distinct systems that conceptually are treated as one system. Unification is about the full integration into one common system or the standardization of information systems with respect to their inner structure, function and content.

Unification as strategy leads to a very high intensity of interoperability or integration. Accordingly, a change in one system must necessarily lead to changes in all other systems. Very often economic and efficiency reasons are the driving force behind unification as strategy. One purpose of unification strategy may be to improve the simplicity, rationality and costs of information management. Another purpose may be to equally treat social events (e.g. pensions, insurances, membership and so forth).

4.2 Intersection Strategy

The second interoperability strategy is Intersection [18][19]. The objective of intersection strategy is to improve the quality of information and the efforts of information management through the elimination of redundancies. Accordingly, the intersection strategy (or overlapping) takes place when the structure of each of the participating systems has one or more elements whose properties are identical (or sufficiently similar) or one of the participating systems has elements which can be used in some of the other systems. Intersection strives to eliminate duplicates.

The intersection strategy creates a shared information space, which can be exemplified by a situation where the participating information systems share some of their constituent parts, i.e. the conceptual base, the information base, or the rule base. In this sense, the participating information systems can be coordinated either (1) by keeping the previously redundant part in one system and allowing the other systems to access to that part, or (2) by unifying the redundant local parts into one shared and common part (like a common database). Thus, a shared information environment might be seen as the result of a coordinating process aiming either to eliminate the overlapping parts of the participating systems or the globalization of some of their parts.

The shared parts become a common property because any kind of change in those parts may have effects on all of the participating systems. Locally, the right to alter the system is limited to unshared parts only. In this sense the effects of integration might be described in terms of expectations for better quality and availability of information services, as well as, in terms of an acceptance of a limited freedom of alteration.

4.3 Interlinking Strategy

Interlinking represents a different concept for systems interoperability. Computerized interaction between different systems is carried out through the exchange of messages which are based on business demands. A significant feature of interlinking is that interoperability takes place without substantial interference with the structures of the participating systems and without substantial limitation of their independence [18][19]. Only information exchange is federated in the interaction agreement, which means that the inner structure in each system can be specified and developed quite independent of the inner structure in all other systems in the total structure. Each system has to automatically produce specified messages according to interaction agreements, either as a planned task or on request, and has to handle incoming specified messages.

The principle has been used in EDI (Electronic Data Interchange) since the 1980's for interaction between information systems in different companies, especially for logistics control, but can be used as a general tool for systems interoperability when there is a high demand for systems independence.

Interlinking implies a change from access or sharing thinking to messaging. If independence is to be preserved, it is not enough to ´open the books´ and to show the data structures in each system. Interaction is instead bridged through the defined messages based on business relations. It is not a question of understanding data in other systems; it is a question of understanding what information is to be transferred between the systems. The different data structures are local and connected to the messages through mapping mechanisms, which must be developed and maintained for each system. If the inner structure in one system is changed, then the mapping mechanism may need to be changed in that very system in order to maintain proper interaction, but there will be no change in any other system as long as the interaction agreement stands.

In the long perspective interlinking creates the possibility to replace a system without any changes in interoperability, as long as the new system fulfills interaction agreement. This has a major impact in that it reduces migration problems and facilitates the sustainability of complex structures of systems.

5 Cases from Swedish Health Care

In this section, we introduce and discuss cases from Swedish Health Care in order to illustrate different interoperability strategies. These cases has been analyzed and discussed during three years with more than 30 professionals from Western Region of Sweden. These individuals currently attend (or have participated) in an IT management master program for professionals. Most of them have worked with IT in the health care area for more than 15 years.

5.1 Background

In Sweden (as in many other countries) health care is decentralized to a number of regional authorities, County Councils. State authorities set the regulations and give directives, but health care is performed in different types of health care units (HCU) in the County Councils and financed locally through taxes in each County.

Within each County a number of private and public health care units may operate. Private units must be authorized by the County Council and must comply with all regulations set by the State Medical Authorities (Ministry of Health and Social Affairs).

For each visit to a health care unit a medical record must be added to the patient's case book by the responsible doctor. When a patient is referred to another health care unit (for instance a specialist clinic), a referral is sent. The response to the referral must always be reported back to the referring doctor, as well as included in the patient's case book. Since a patient is free to book a visit to a different HCU (also in a different County) the patient may have a number of different medical case books in different health care units in different Counties. Emergency treatment and the fact that patients move result in patient history being split among several case books.

If it was possible for the treating doctor to get relevant information from different case books, treatment effectiveness would improve, costs will be reduced and better quality could be given in health care. The minister of Health and Social Affairs has stated that this is an issue of utmost importance, and has instructed the Swedish Health Care to meet this demand. Solutions must be nationwide and consider interaction with private HCU:s as well as possible changes in healthcare organization.

IT support for medical records

The amount of information in medical case books is huge and calls for IT solutions. The acquisition of information systems to handle medical records can however so far be characterized as somewhat unstructured. Different health care units have acquired their own systems, sometimes based on initiatives from interested doctors. There are two main reasons for this decentralized or even anarchistic approach: a legal reason in that the unit is responsible for medical records within that unit and a professional reason, in that a certain medical discipline may have special demands, which are not relevant for other disciplines. The result is a diversity of systems from many different vendors resulting in a very low degree of standardization (low level of unification). In addition, even the same type of health care units may have different types of systems implemented (e.g. within radiology). The systems are generally well adapted for medical demands within the actual healthcare unit, and they also fulfill demands on privacy and integrity as all patients records are kept within the healthcare unit. A main problem is, however, that the computerized interaction between these systems is very limited. For referrals between health care units and support units (for instance blood test and bacteriological tests) some communication solutions have been developed and implemented. Transfer of medical records from one unit to another is however still performed manually in most cases.

The challenge is now to create architectural solutions for systems interoperability that are in harmony with demands from healthcare, both in a short (business agility) and long term perspective (strategic alignment).

The problem of "need to know"

For each health care unit the IT support may be sufficient, but from an overall perspective the problem is that information from a medical record in one system is not available for treating doctors in other healthcare units. The solution to the problem can be divided into two tasks:

a) knowing in which other units treatment has been (or is) carried out and in which systems medical records are stored, and
b) getting relevant information out of these systems.

Evidently there is a need for interoperability between different information systems for medical records in order to support these two tasks. When seeking solutions for this some conditions must be considered:

- Only medical records from relevant types of visits to other healthcare units should be available. If, for instance, the actual concerns a leg fracture, the doctor should not be allowed to read medical records on the patient's treatment for drug addiction.
- The doctor must have the patient's direct permission to read and use medical records from other healthcare units.
- If the doctor receives a medical record from some other healthcare unit, this record must be inserted in the actual case book.

5.2 Alternative for Interoperability Based on Unification Strategy

A seemingly simple solution to the problem is to create a common system where all medical records are stored. Just a single search on an identified patient would make the total computerized case book directly available. This is an application of the unification strategy based on the "one common system" principle. This alternative means that single set of general medical records must be defined, which are relevant and useful for all medical disciplines and for all different types of treatment. Furthermore, this central system must fulfill all legal demands on integrity and patients´ rights. Finally, the system must gain general acceptance as it is either to replace existing systems or will have a major impact on current systems in use.

Projects based on this unification principle have been started at different levels. An observation is that in project proposals no other strategy are evaluated. Project managers seem to have the opinion that there is no other real alternative to unification. Results have been reached, for instance using a central system in a hospital, but attempts to develop a total system at regional level have so far failed [20] [21]. A project for a common regional system for medical records was in 2008 terminated due to:

- too long time for development and increasing costs, and
- specified functionality could not satisfy the demands from all the different interests.

Furthermore, the system and its resultant organizational impact were not accepted out in the organization. Even if regional systems were possible, interaction between these regional systems at national level would still be an unsolved problem. As different County Councils develop their own solutions independent of each other, one single total national system seems destined to fail.

5.3 Alternative for Interoperability Based on Intersection Strategy

The intersection strategy for interoperability should mean unification of only those parts of data storage which are of interest for interoperation, allowing access to other systems for retrieval on demand. There are groups of information systems for medical records that are homogenous, as they emanate from the same standardized system package from a certain vendor. Within each of these groups it could be possible to apply the intersection principle, in order to solve task b) (the getting problem) specified above, without major changes in existing systems. In total there is, however, a high degree of heterogeneity in the total structure of systems. To implement access ability and intersection for heterogeneous systems calls for major changes in or replacement of all existing systems, and then we are back to the same problems that are listed for the unification strategy.

In a longer perspective, intersection will cause maintenance problems. A change or replacement of a system which includes changes in the inner structure for that system will have an impact on all other system in the structure, as coordination of inner structure and data storage is the basis for interoperation in order to facilitate access to stored data.

For the “need to know” problem in health care the intersection principle does not automatically solve the problem of knowing where the relevant information resides. Thus, there is a need for some type of common system in order to make known in which of the other systems relevant medical records for the actual patient may exist.

5.4 Alternative for Interoperability Based on Interlinking Strategy

As this alternative is not quite obvious, a more detailed analysis may be necessary in order to explain differences and possible outcomes from the alternative.

The first task a), knowing where relevant information resides, calls for some kind of medical visit directory. This is sometimes referred to as RLS, Record Locater Service. In figure 1, a special system MED DIR is inserted for this purpose. In an extended structure it is possible to have more than one directory system.

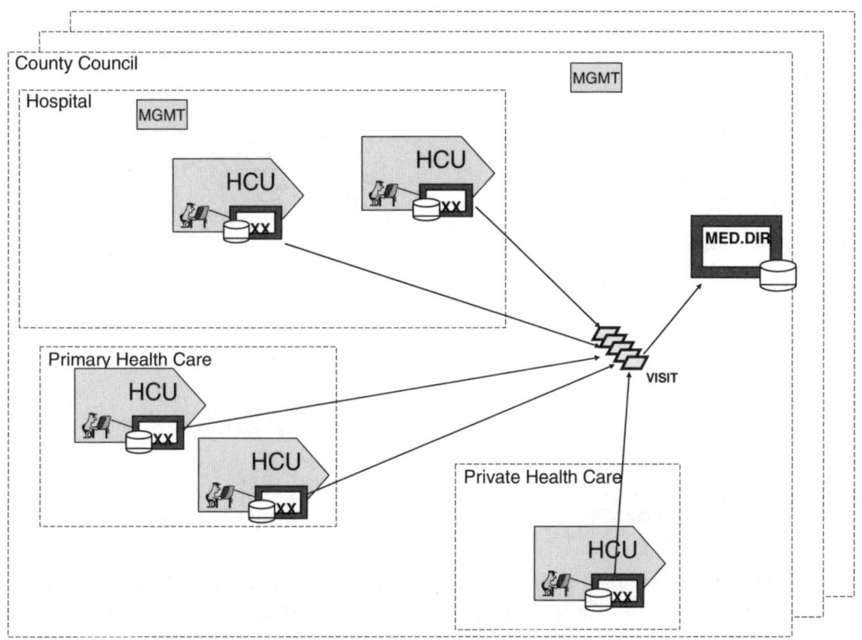

Fig. 1. Medical Directory, interacting with local systems applying interlinking

A message (VISIT in fig1) is defined for reporting each medical visit. In order to solve task b, the message must identify the patient and indicate the type of visit in order to facilitate the choice of relevant system for later information exchange. All attached systems must produce this type of message, and medical visits are thus automatically reported to MED.DIR through this planned messaging. The patient has however the right to decide that a certain medical record should not be available out of the treating healthcare unit. If so, the VISIT record is not sent. Existing systems (and additional systems that gradually will enter the structure) must be given the functionality to create this message, but the inner structure for data storage in each system is not affected. Changes or replacement of a certain systems (including MED.DIR) can therefore be carried out without any changes in other systems as long as the message VISIT can be sent/received. The solution for task a) can thus gradually be established, and can be maintained in a long term perspective, without major changing the inner structure for existing systems.

The second task b), making relevant information from other systems available for the treating doctor can now be supported the following way: (see fig 2).

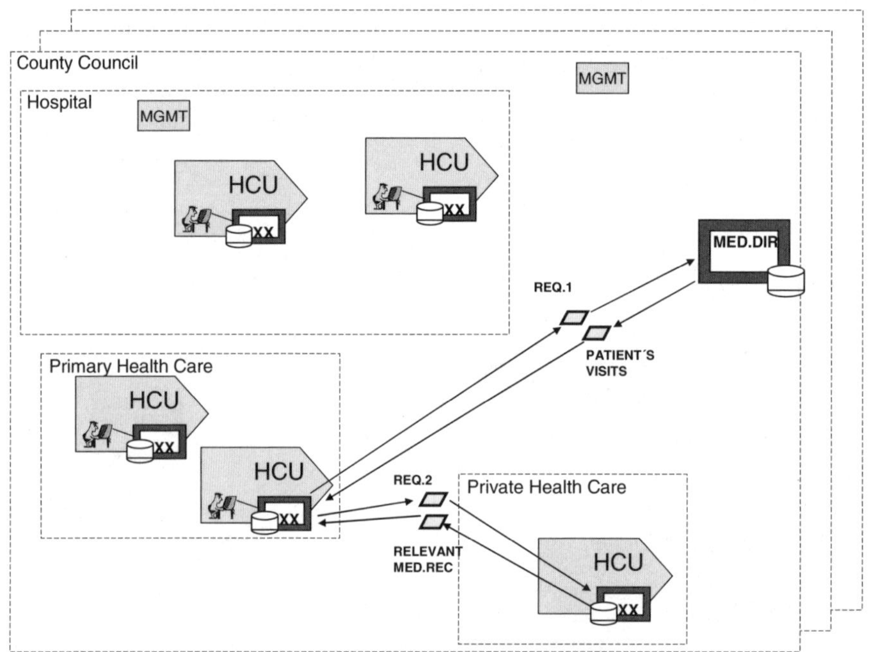

Fig. 2. Interlinking strategy for transfer of relevant Medical Record

A request is sent to MED.DIR from the system that supports the treating doctor. This message (REQ.1 in fig2) is basically patient's identification. The automatic answer to this request is a message pointing out in which systems the actual patient has medical records. The treating doctor has to judge which records can be relevant, and (after authorization and permission from patient if possible) to decide on which records to pick up. The transfer of medical records can be achieved by using Request – Answer messages (applying the interlinking principle). Both request and answer must be centrally defined, and this is a critical task. The main objective is to make the record available and readable for the treating doctor. However, a number of varieties of medical records may exist because the different medical disciplines. The point is however, that only interaction must be centrally defined. The inner structure in each system must not be known outside the system, and the retrieval from local storage is performed inside the system, in order to produce the relevant message as an answer to the request. The received relevant medical records from other systems are locally inserted in the actual patient's case book according to the local systems storing principles. It is essential that information transferred from other medical units is inserted without any changes.

The interlinking strategy allows establishing of interoperability for medical records step by step. When MED.DIR is implemented different local systems can gradually be attached. The task of "knowing where" (task a) can thus be incrementally supported, even if the computerized solution for the task of "getting relevant information" (task b) is not yet fully implemented, i.e. the transfer of medical records must be carried out manually. The interlinking alternative for systems interoperation maintains independence between the systems. This means in a long term perspective that this strategy can support both a) and b) through a sustainable structure in which each system can be changed or even replaced without consequences for any other systems, given that the new /changed system can receive and send specified messages.

6 Summary and Conclusion

The focus in this paper has been on elucidation of the relation between interoperability strategies, business agility and alignment. Our main point is that an interoperability strategy has significant impact on business agility and should therefore be careful chosen in order to avoid misalignment. This decision must be based on an analysis of the pros and cons of various alternatives in relation to the requirements from the business. Our experience is that this kind of analysis is seldom done in practice (which is validated from the health care cases).

Business agility, in general, may be seen as a capability to respond to changes and the ability to change in a purposeful way rather than in an ad hoc manner. However business agility presupposes information systems interoperability. The larger, heterogeneous, and dynamic the business and its environment are, the more crucial are the issues of information systems interoperability. However, the information systems interoperability does not live in its own world; it must be aligned with business requirements.

The concept of interoperability in general refers to the ability of two or more business units to exchange or share information and in our case especially through the use of their information systems. Accordingly, the nature of the business environment should define the pattern or strategy upon which the business entities and their information systems should interoperate. Thus, in situations where the environment is characterized by homogeneous and relatively stable requirements the business units may follow global unified rules and principles and the ability to share information may increase. Accordingly, systems interoperability may follow the same unified rules and sharing principles. However, if the situation is characterized by heterogeneous and dynamic requirements, the involved business units may emphasis ability to change and thus also ask for similar ability concerning their information systems. Interoperability in general and systems interoperability in particular presupposes a well established alignment between business units and their corresponding information systems.

Alignment is any form of fitness or workable harmony between the requisites of business units and the capabilities of information systems to satisfy these requirements. Perfect alignment is a utopia because every business enterprise exists in a continuum of changes occurred in the domains of business, domain of information systems as well as in the domain of technology. Furthermore, there are no organizations that are 100% agile (at least to our knowledge). Large organizations have areas where agility is required, and areas where agility is not crucial. All three interoperability strategies may be relevant, but in different situations. This must be taken into account in the broader perspective of enterprise architecture.

Enterprise Architecture is an expression of workable alignment between business requirements and systems assets that have been organized in a particular way as a response to the nature of business environment. Therefore, Enterprise Architecture Frameworks should give guidelines and principles for the choice of interoperability strategy. Such guidelines and principles should take into account:

- How to delineate information systems and create "systems of systems".
- How the different systems should interoperate in order to satisfy the expectations of the business and its environment.
- How to manage the requisites of alignment between business architecture and information systems architecture.

Unfortunately, very few existing Enterprise Architecture Frameworks give clear guidance about how to architect and manage the critical issues of interoperability. Thus, there is a need for more knowledge about enterprise architectures that holds in harmony 1) the ever changing nature of business and 2) the capabilities and assets to respond quickly to these changes.

In this study, we argue that the choice of interoperability strategy has a significant impact on business agility and should therefore be analyzed and evaluated before launching development projects for systems interoperability. The choice of a suitable strategy is a main condition for aligning information systems interoperability to business demands and to create agility. If the wrong strategy is chosen, there is a considerable risk for misalignment and expensive consequences.

References

1. Benson, S., Dove, R., Kann, J.: An Agile Systems Framework: A Foundation Tool. In: Proceedings of Annual Conference Agility Forum (1992)
2. Baskerville, R., Mathiassen, L., Pries-Heje, J., De Gross, J. (eds.): Business Agility and Information Technology Diffusion. Springer, NY (2005)
3. Christopher, M., Towill, D.: Supply Chain Migration from Lean and Functional to Agile and Customised. Supply Chain Management 5(4), 206–213 (2000)
4. Yusuf, Y.Y., Sarhadi, M., Gunasekaran, A.: Agile manufacturing: the drivers, concepts and attributes. International Journal of Production Economics 62(1/2), 33–43 (1999)
5. Alberts, D.S., Hayes, R.E.: Understanding Command and Control. CCRP Publication Series (2006)
6. Ramasesh, R., Kulkarni, S., Jayakumar, M.: Agility in manufacturing systems: an exploratory modelling framework and simulation. Integrated manufacturing Systems 12(7), 534–548 (2001)

7. Kidd, P.T.: Agile Manufacturing, Forging New Frontiers. Addison-Wesley, London (1995)
8. Holmqvist, M., Pessi, K.: Agility through scenario development and continuous implementation: A global aftermarket logistics case. In European Journal on Information Systems, special issue on Business Agility and IT Diffusion (2006)
9. Zachman, J.A.: A Framework for Information Systems Architecture. IBM Systems Journal 26(3), 276–292 (1987)
10. Sowa, J.F., Zachman, J.A.: Extending and Formalizing the Framework for Information Systems Architecture. IBM Systems Journal 31(3), 590–616 (1992)
11. The Open Group: The Open Group Architecture Framework: Version 8.1.1, Enterprise Edition (2007)
12. CIO Council: A Practical Guide to Federal Enterprise Architecture. Chief Information Officer Council, Version 1.0, February 2001 (2001)
13. Aerts, A.T.M., Goossenaerts, J.B.M., Hammer, D.K., Wortmann, J.C.: Architectures in context: on the evolution of business, application software, and ICT platform architectures. Information & Management 41, 781–794 (2004)
14. van der Poel, P., van Waes, R.: Framework for Architectures in Information Planning. In: Falkenberg, I.E., Lindgreen, P. (eds.) Information Systems Concept: An In-depth Analysis, Elsevier Science Publishers, North Holland (1989)
15. IEEE STD 610.12. Standard Glossary of Software Engineering Terminology, IEEE (May 1990) ISBN: 155937067X
16. Clark, T., Jones, R.: Organisational Interoperability Maturity Model for C2 (1999), http://www.dodccrp.org/events/1999_CCRTS/pdf_files/track_5/049clark.pdf
17. Alter, S.: Information Systems. The Benjamin/Cummings Publishing Company, Inc. (1996)
18. Solotruk, M., Kristofic, M.: Increasing the Degree of Information System Integration and Developing an Integrated Information System. Information & Management 3(3) (1980)
19. Magoulas, T., Pessi, K.: Strategic IT Management. Doctoral Dissertation Department of Informatics Gothenburg, Sweden (in Swedish) (1998)
20. Bäck, M.: Missarna som knäckte GVD (in Swedish) (2007), http://itivarden.idg.se/2.2898/1.130869
21. Jerräng, M.: Fiaskot kostar 300 miljoner (in Swedish) (2008), http://itivarden.idg.se/2.2898/1.147465

Towards a Business-Oriented Specification for Services

Linda Terlouw[1,2]

[1] Delft University of Technology,
Mekelweg 4, 2628 CD Delft, The Netherlands
l.i.terlouw@tudelft.nl
[2] Ordina,
Ringwade 1, 3439 LM Nieuwegein, The Netherlands

Abstract. By far the best known standard for registering and searching for services is the UDDI. A great weakness of this standard is its technology-driven way of specifying services; it is still inadequate for specifying the majority of aspects that are relevant from a business point of view. This stands in sharp contrast to the main premises of SOA, i.e. increased flexibility by the reuse of services and better business/IT-alignment by speaking the same language. A more comprehensive approach to specifying services is the business component specification framework. One of the aspects that needs to be specified according to this framework are the business tasks. The framework, however, does not define precisely what a task is and how a task should be identified. In this paper we propose taking the enterprise ontology as a starting point for specifying these tasks. Furthermore, we demonstrate our approach using a life insurance company case.

Keywords: SOA, service specification, enterprise ontology.

1 Introduction

As *Service Oriented Architecture* (SOA) and *Service Oriented Design* (SoD) are gaining popularity in industry, *enterprises*[1] struggle with the question how to *identify* and *specify* services that support the execution of their *business processes.* Several approaches for the identification of services exist, for instance business process decomposition [1], component-based identification [2,3], and legacy system analysis [4]. Once identified, services need to be specified in order to (i) implement them, either by building new software systems or by using existing ones and (ii) enable potential service consumers to find the services they require. *Service registries* act as a means for storing these service specifications.

When looking into the contents of service registries of large enterprises, one often finds many technical services that have little meaning and relevance to business people. Also, services that are of interest to business people are usually

[1] By enterprise we mean commercial as well as non-profit organizations as well as networks of organizations.

J.L.G. Dietz et al. (Eds.): CIAO! 2008 and EOMAS 2008, LNBIP 10, pp. 122–136, 2008.

specified in a way that is incomprehensible for them, because they simply do not 'speak' the Web Service Description Language (WSDL) [5]. Even though business people do not have to browse through the service registry themselves, they do need to validate whether a certain service really fits their needs for business process support.

In this article we build on the work on enterprise ontology [6] and business components specification [7]. More precisely formulated, we take the Actor Transaction Diagram and the Object Fact Diagram of the enterprise ontology as a starting point for the specification of the business tasks a service supports. The main contribution of work lies in the field of service specification; the relationship between the service and the enterprise ontology makes it clear to a business analyst what business value the service brings. Though a more fundamental problem remains, i.e. the gap between formal, unambiguous models needed for specification and the natural language that is best understood by business analysts, our work is a first step in speaking in real business terminology instead of technical interface descriptions.

The rest of this paper is organized as follows. Section 2 provides an overview of the current approaches to service specification, including the business component framework. In section 3 we introduce a template based on the enterprise ontology for specifying the business tasks a service supports. In section 4 we discuss an insurance company case to show the practical value of our work. Finally, section 5 concludes the paper by summarizing how we contributed to the service specification process.

2 Current Approaches to Service Specification

Before we provide our own approach in section 3, we discuss related existing work on service specification in the next paragraphs.

2.1 Specification Using the UDDI

Universal Description Discovery and Integration (UDDI) [8] is the web service standard for publishing and discovering services. The UDDI distinguishes between the following specification elements: businessEntity, businessService, bindingTemplate, and tModel. Each businessEntity contains information about a service provider such as the name of the provider and the details of the contact person. This businessEntity provides zero or more businessServices. The businessService is the actual service specification which consists of a name and a description in free-format natural language. If this service is implemented, then it has a reference to a interface description file, or WSDL file, specified in the bindingTemplate. The bindingTemplate in its turn can reference to a tModel. This tModel is not very precisely defined in the UDDI standard and can specify among others industry taxonomies, service categories, and technical specifications.

OASIS provides an overview of UDDI related products [9]. Most vendors of these products are also vendors of Enterprise Service Bus products and have a focus on the run-time environment.

2.2 Specification Using Semantic Web Services

In order to make service registries more powerful the UDDI standard is sometimes combined with semantic web service standards like OWL-S [10,11], WSMO [12], and WSDL-S [13]. Also more generic approaches for semantically enhancing UDDI registries are possible [14]. The goal of semantic web services is thoroughly specifying every aspect of a service in order to enable automatic matching of supply of and demand for services. It takes a lot of effort (if feasible at all) for a large enterprise to specify everything into so much detail that automatic matching on run-time becomes possible.

At this moment none of the semantic web service approaches are popular in industry. Though industrial partners participate in research projects, we see little (if any) semantic service registries in real SOA environments. In practice supply and demand is still matched by a human being.

Until recently semantic web service approaches did not make a clear distinction between the specification of the service itself and the specification of a task a service supports. Both were regarded at the same level of granularity. Recent work [15] builds on WSMO to create more clarity on how semantic web service support business processes by introducing the standard BPMO. The meta-model of BPMO specifies among others business goals, business processes, business roles, business rules, and business data.

2.3 Specification Using Business Component Specification

A business component is a software component of an information system that support directly the activities of an enterprise. These business components are reusable, self-contained and marketable [7] and are accessed via services. Even though business components are always accessed by services, services do not have to be constructed of components, but can just as well be constructed of, for example, a monolithic system. In SOA we simply don't care about the construction and implementation.

Figure 1 shows the specification aspects for a business component. A more thorough explanation is provided in the memorandum "Standardized Specification of Business Components" [7]. A prototype of a tool for specifying business components is provided by Zaha and Albani [16]. As far as we know, no commercial versions of this kind of registries are available.

2.4 Evaluation

In Table 1 we provide an overview of the requirements to which a service specification needs to conform to make it understandable and useful to a business analyst. These requirements are based on experience in several case studies. We describe the consequences of not conforming to these requirements to illustrate their importance.

When we recall the different approaches described in this section, we see that none of them actually conforms to all requirements (see Table 2).

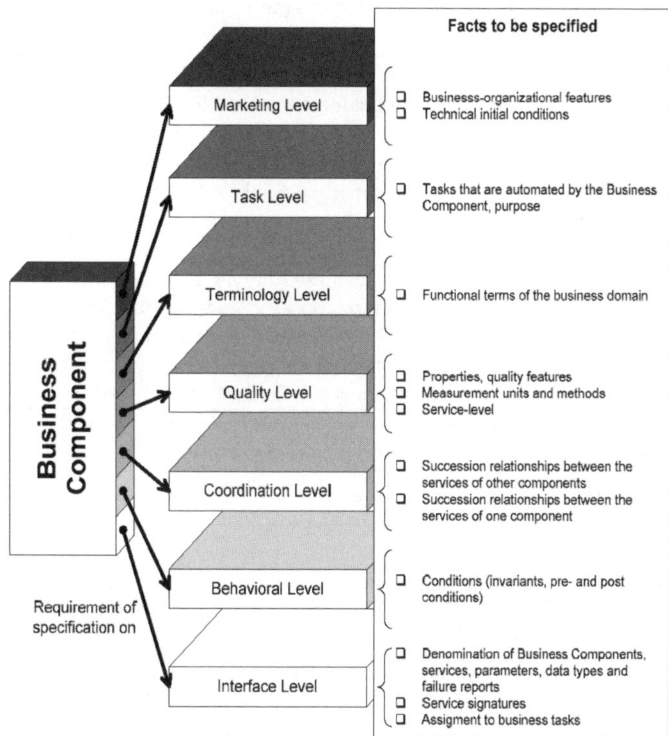

Fig. 1. The business component specification framework [7]

The UDDI conforms to none of the requirements since it is a very technical oriented standard. Although the concepts BusinessEntity, BusinessService, and BindingTemplate are relatively well defined, the tModel is used (misused?) in very different ways. Though the tModel can be used for the categorization of services, it remains unclear how services should be categorized and how the categories relate to business tasks.

Most semantic web services standards describe the function that a service fulfills in a consistent way and process models describe the relationship between services. The semantic web service approaches make a distinction between the service itself and the business task that the service supports, however, they are treated at the same level of granularity. In order to make the semantic models understandable to a human business analyst, we need to answer the questions: what is really relevant from a business perspective and what can be omitted and how do we describe tasks in such a way that they are stable?

The business component framework clearly distinguishes between the service itself and the business task that is fulfilled. The function of the service is specified in terms of its interface and behavior. The task that a service supports is described in business terminology. An example specification of the tasks supported

Table 1. Requirements for business tasks specification

Nr.	Requirement	Consequence of not conforming
r1	It is clear which business task(s) a service supports	The business analyst does not know what the business value of a service is and he cannot judge whether the service fits his purpose or not.
r2	Relationships between tasks are clearly described	The business analyst loses overview of the total picture and searching for service can become time consuming.
r3	Task are formulated in a consistent way	The business analyst may not find the services that fit his purpose or may find services that not find his purpose, because he is not sure whether the task description of the service provider is consistent with his own view on the task description.
r4	The tasks are stable, i.e. tasks descriptions only change when the essence of the enterprise changes	The business analyst needs to deal with constantly changing task descriptions which is time consuming.

Table 2. Evaluation of specification methods from business point of view

Nr.	UDDI	Semantic WS	Business Component
r1	possible to specify using the tModel	most approaches describe business tasks or activities at the same level of granularity as the services, except for BPMO	tasks are acknowledged as one of the main specification aspects for a service
r2	relationships between tasks cannot be specified	relationships between tasks are specified using logic	relationships between tasks are specified using reconstructed functional language or natural language
r3	no consistently applied structure	tasks are specified using logic	tasks are specified in reconstructed functional language or natural language
r4	not clear how to identify stable tasks, tasks are in general quite low level and therefore unstable	not clear how to identify stable tasks, tasks are in general quite low level and therefore unstable	not clear how to identify stable tasks, tasks are in general quite low level and therefore unstable

by the (services of the) business component, as depicted in Table 3, is provided by Fettke and Loos [17]. Though the authors specify these tasks in business terminology, the tasks are still quite low level and little insight is provided in why we need to look up a bank code or how this task is related to other tasks. Since the business component framework does not prescribe how to identify the tasks, one does not know whether tasks are stable or not.

Table 3. Example task specification [17]

Task name	Task description
verify bank code	this task verifies if a given bank code is valid or if a given bank code corresponds to a given bank name.
look up bank code	this task looks up the bank code for a given bank name.
look up bank	this task looks up the bank name for a given bank code.

3 Specifying Business Tasks Using the Enterprise Ontology

In this section we describe how we use the enterprise ontology as a starting point for specifying which tasks a service supports. We propose to take the enterprise ontology as a basis for specifying which business tasks a service supports, because it has two main advantages over most other business modeling methods; (i) the models are quite compact, because non-ontological transactions are omitted and coordination acts are captured in the transaction and (ii) different modelers create the same models because the enterprise ontology is more than only a modeling language; it is also a way of thinking.

3.1 Definitions

We first need to clarify what we mean by a service as a lot of different interpretations exist [18]. We define a service as follows based on the ideas of the enterprise ontology:

Definition 1. *A service is a task offered by a service provider to (potential) service consumers that conforms to the following properties:*

1. *it is accessible through an interface;*
2. *it is described by a service specification that provides information for the service consumer to find and use the service;*
3. *its implementation is hidden to (potential) service consumers;*
4. *it is autonomous, i.e. it is designed, deployed, versioned, and managed independently of other services;*
5. *it supports a transaction;*
6. *it is either ontological, infological, or datalogical.*

The first four properties are often encountered in definitions of services. The last two properties require some clarification. Dietz [6] defines in the second axiom of his Ψ-theory that *coordination acts* are performed as steps in universal patterns. These patterns, also called *transactions*, always involve two *actor roles* and are aimed at achieving a particular result. The basic transaction pattern consists of the following coordination acts:

- *request*: the initiating actor role request the executing actor role to perform a certain production act;

- *promise*: the executing actor role promises to execute the production act;
- *state*: the executing actor role states that the production act has been executed;
- *accept*: the initiating actor role accepts the result of the production act.

The basic transaction pattern can be extended with dissent patterns and cancellation patterns to represent more complex coordination mechanisms. The enterprise ontology only takes into account transactions at the *ontological* level, where ontology is defined as follows [6]:

Definition 2. *the ontology of a system*[2] *is the understanding of the system's operation that is fully independent of the way in which it is or might be implemented.*

What this means is that the enterprise ontology only includes transactions that are concerned with the bringing about of new, original thing like decisions, judgments etc. Therefore the ontological business process model is quite stable and business processes from enterprises in the same industry tend to look very much the same.

Besides directly supporting ontological transactions, services can also support transactions by providing infological and datalogical capabilities. Infological services deal with the processing of information. Datalogical services deal with the recording of and the transportation of recorded information items.

In the field of SOA, most architects use the words 'orchestration' and 'business process' as synonyms. We, however, make a clear distinction between the orchestration, the complete business process, and the ontological business process.

An *orchestration* controls the sequence of service calls. This orchestration has a graphical representation (flowchart) and machine-readable representation (usually in XML). When executing, the orchestration (i) calls automated functionality by invoking automated services (ii) coordinates human services by taking input from and providing output to a human user through a portal. When an orchestration only invokes automated services and does not need any manual intervention, it is said to support *straight through processing.* Manual services in an orchestration are called *human workflow.* Human workflow is increasing in popularity at this moment and the main standard for orchestration, Business Process Execution Language (BPEL), that originally was meant only for orchestrating software services, was extended with BPEL4People [19] in 2007.

The *complete business process* consists of all the activities that need to be performed to reach a certain business result. For the specification of stable business tasks, we are only interested in the *ontological business process* which is defined as follows [6]:

Definition 3. *A collection of causally related transaction types, such that the starting step is either a request performed by an actor role in the environment or a request by an internal actor role to itself.*

---

[2] The notion of system is used in a broad meaning and does not only apply to automated systems.

Table 4. Types of processes and their scopes

Type of process		Scope of process
Ontological business process		Controls the sequence of ontological transactions
Complete business processes		Controls the sequence of all business activities
Orchestration	Straight through processing	Controls the sequence of automated services
	Human workflow	Controls the sequence of (messages to and from humans for coordinating) human services

Figure 4 exhibits the different types of process we mentioned in this paragraph and their scopes.

3.2 The Specification Template

Figure 5 exhibits our template for the specification of the business tasks that a service supports. In the template we specify the service layer and category, the transaction type(s) that a service supports, and the object classes(es) from the result type of the transaction type(s). The transaction types and object classes are directly available from the enterprise ontology. The service layer and service category require some more explanation.

Although, according to the Ψ-theory, always a human actor is responsible and has authority to fulfill a certain actor role, information technology can support human subjects in the coordination and execution of transactions. The coordination steps of a transaction can be automated using human workflow. The execution step of the transaction is not always automatable. In case of transactions that affect concrete objects, like the preparation of a pizza, one only can

Table 5. Template for the specification of the business tasks

<table>
<tr><th colspan="2">Template for task specification</th></tr>
<tr><td>Service layer:</td><td><ontological (B) | infological (I) | datalogical (D)></td></tr>
<tr><td>Service category:</td><td><calculation service | validation service | selection service | matching service | registration service | retrieval service...></td></tr>
<tr><td>Supported transaction:</td><td>The transaction that the service supports</td></tr>
<tr><td>Description of transaction:</td><td>Explanation of the transaction in natural language</td></tr>
<tr><td>Result of transaction:</td><td>The result of the transaction that the service supports</td></tr>
<tr><td>Executor of transaction:</td><td>The executor of the transaction that the service supports</td></tr>
<tr><td>Initiators of transaction:</td><td>The initiator(s) of the transaction that the service supports</td></tr>
<tr><td>Related transactions:</td><td>Transaction that are one step away from the transaction that the service supports</td></tr>
<tr><td>Object class:</td><td>The object class that is referred to by the result type of the transaction.</td></tr>
<tr><td>Related object classes:</td><td>The object classes related to the previously mentioned object class.</td></tr>
</table>

automate the administrative reflection of the status of the real world. In this situation the only possible automated service is the administration of the P-fact "pizza x has been prepared on <date> <time>". In case of transaction that do not affect concrete objects, one can also automate the execution itself. Take, for example, the quotation of an insurance policy. We call services that directly support the execution step of ontological transactions business services. These services cannot be divided into categories since they are specific to a certain industry.

Infological services deal with the processing of information. Looking at the infological services, we can find a subdivision of functionality in among others *calculation services*, *validation services*, *selection services*, and *matching services.* An example of a calculation service is the calculation of an insurance premium for a specific person. A validation service checks whether a certain conditions holds, e.g. checking whether a potential client is not on a black-list. A selection service makes it possible to select one or more values from a list of values. Matching certain values, like received payment on a bank account and outstanding accounts, can be done with matching services.

Datalogical services deal with the recording of and the transportation of recorded information items. If the recording of information items is automated, then the services used for recording are analogue to create, read, update and delete (CRUD) functions for databases. We prefer, however, not to use the term CRUD, because the words that form this abbreviation have no meaning to business analysts and suggest that one is working on database schemas instead of conceptual business objects that may be spread over multiple databases. Rather we speak of *registration, retrieval, alteration, and removal services.* If the transportation of recorded information items is automated, then we require *distribution services*, e.g. an insurance agent passes information about a new client to an insurer. If the transportation of information items is not automated, then services that transfer data from and to the real world could be used: *virtual-to-real-world services* and *real-to-virtual-world services.* The first type of service, e.g. a service for printing, is more common and more easy to implement then the second one, e.g. a service for retrieving sensor information.

Using the enterprise ontology we can make services on a lower level traceable to essential business transactions. Either directly or indirectly services support one or more transactions. The purpose of a service is therefore always clear.

4 The Insurance Company

4.1 Background

The insurance company Protector provides three types of *life insurance* products; term life insurance, pension insurance, and capital sum insurance. For each type of product the life insurance policy has an insurant, one or more insured, and one or more beneficiaries. The insurant is an organization that or person who is responsible for the payment of the premium of a policy. The insured is a person

Table 6. Transaction Result Table (TRT)

	Transaction type	**Result type**
T01	product advising	product advice *adv* is created
T04	policy quotation	policy *pol* is quoted (only for individual policies)
T05	policy binding	policy *pol* is bound
T06	premium payment	premium is paid for policy *pol* for premium period *per*
T07	voluntary deposit	voluntary deposit is made for policy *pol*
T17	reinsurance of policies	policy collection *pco* is reinsured for per
T18	reinsurance premium payment	reinsurance premium is paid for policy collection *pco* for period *per*
T26	commission payment	commission *com* is paid
T27	policy underwriting	underwriting for policy *pol* has been done

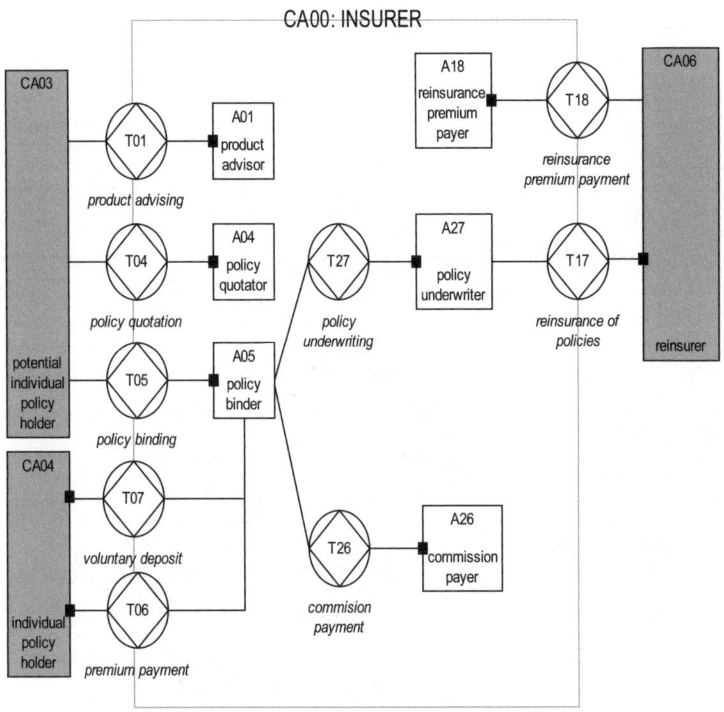

Fig. 2. Actor Transaction Diagram (ATD)

who is the 'insured object'. The beneficiary is a person who receives a payment if the insurant has right to a benefit according to the product rules of a policy.

Protector sells the products either to a company, i.e. *collectively*, or to an individual person, i.e. *individually*. Some products may be sold collectively as well as individually, some only collectively or individually. An example of a collective insurance is a pension insurance provided by a company to its employees. Usually, an employee can choose whether or not he participates. An example of an individual insurance is a term life insurance related to a mortgage. We use the word 'policy' for the individual policy as well as for a contract participation.

In the next sections we will show some of the models of the enterprise ontology of Protector and we will apply our template to two example services.

4.2 Dealing with New Individual Policies

Table 6 exhibits the subset of the transactions of the life insurer that is of relevance for handling new individual policies. Figure 2 depicts these transactions graphically in an Actor Transaction Diagram. The appendix presents the Object Fact Diagram for the life insurance company in an ORM-based language. The result types of Table 6 are depicted as diamonds in this diagram.

Table 7. Business task specification of CalculatePremium service

CalculatePremium Service	
Type of service:	infological
Subtype of service:	calculation service
Supported transaction:	T04: policy quotation
Description of transaction:	offering a policy of a product to a potential policy holder
Result of transaction:	R04: policy *pol* is quoted
Executor of transaction:	A04: policy quotator
Initiator of transaction:	CA03: potential individual policy holder
Related transactions:	none
Supported transaction:	T27: policy underwriting
Description of transaction:	determining whether a potential individual policy holder is accepted as insured or not and on what terms
Result of transaction:	R27: underwriting for policy *pol* has been done
Executor of transaction:	A27: policy underwriter
Initiator of transaction:	A05: policy binder
Related transactions:	T05: policy binding, T17: reinsurance of policy
Object class:	POLICY
Related object classes:	POLICY *pol* belongs to POLICY COLLECTION *pco*
	the payer of INSURANCE PREMIUM *pre* is the insurant of POLICY *pol*
	the beneficiary of INSURANCE BENEFIT *ben* is the beneficiary of POLICY *pol*
	PERSON *per* is the beneficiary of POLICY *pol*
	PERSON *per* is the insured of POLICY *pol*
	PARTY *par* is the insurant of POLICY *pol*
	PRODUCT *pro* is the product of POLICY *pol*

The process of handling new individual policies runs as follows[3]. A potential individual policy holder requests a quotation for a product with or without getting an advice first. After the promise of the policy binder, the policy underwriting is

[3] The graphical representation of the process, the process step diagram, is omitted due to space limitations.

Table 8. Business task specification of RegisterAdvice service

RegisterAdvice Service	
Type of service:	datalogical
Subtype of service:	registration service
Supported transaction:	T01: product advising
Description of transaction:	giving advice on the product that bests suits the potential individual policy holder needs
Result of transaction:	R01: PRODUCT ADVICE adv is given
Executor of transaction:	A01: product advisor
Initiators of transaction:	CA03: potential individual policy holder
Related transactions:	none
Object class:	PRODUCT ADVICE
Related object classes:	the advised PRODUCT in PRODUCT ADVICE *adv* is *pro*
	the creator of PRODUCT ADVICE *adv* is AGENT *age*

requested. The policy underwriter checks if the risk is acceptable and optionally request reinsurance of the policy. Reinsurance is also known as 'the insurance for insurance companies'. It protects an insurance company against exposure to large risks. Sometimes regular insurance companies fulfill the role of reinsurer, sometimes specialized companies fulfill this role. If (i) reinsurance is necessary and the reinsurance states or (ii) if no reinsurance is needed, then the underwriter states. The policy binder accepts and then request premium payment to the insurant and optionally requests commission payment to the agent.

4.3 Example Services

We applied our template to two services of Protector: the CalculatePremium service and the RegisterAdvice service. Tables 7 and 8 depict their specifications. The first service calculates the premium for an individual policy, which is used for supporting transactions T04 and T27.

The second service stores the advice that a product advisor has given to the potential individual policy holder and supports T01.

5 Conclusions

Although the UDDI is currently the most popular standard for implementing service registries, it lacks the ability to specify services from a business point of view. Its original intent was to provide a world-wide registry for accessing services at run-time. In practice, however, most service registries are used within one organization or organizational network and services are found on design-time. Even for this purpose the information in the UDDI is too limited to determine whether or not the service is useful for supporting a specific business process. Still a lot of work on service specification needs to be done.

In this paper we built upon the work on business component specification and formal enterprise modeling. The business component specification framework prescribes seven levels that need to be specified. Though standards exist for the lower-level layer of the business component specification framework, e.g. WSDL for the interface level and UML Object Constraint Language (OCL) for the behavioral level, there are less formal models for the higher levels, e.g. the terminology, marketing, and task level. We have provided a contribution to the specification of the task level, which specifies which business tasks are supported. We have used the enterprise ontology as a basis for a business task specification template. This template makes services traceable to the ontological transactions they support, i.e. their business purpose. We have shown the practical relevance of this work in a life insurance company case study of which we have discussed parts in this paper.

Future research will address the design of an XML-based standard for the proposed template and its integration with the UDDI standard.

References

1. Henkel, J.Z.M., Johannesson, P.: Service-based processes: design for business and technology. In: ICSOC 2004: Proceedings of the 2nd international conference on Service oriented computing, pp. 21–29. ACM Press, New York (2004)
2. Albani, A., Dietz, J.: The benefit of enterprise ontology in identifying business components. In: WCC 2006: Proceedings of the IFIP World Computer Congress, Santiago de Chile, Chile (2006)
3. McGovern, J., Sims, O., Jain, A., Little, M.: Enterprise Service Oriented Architectures: Concepts, Challenges, Recommendations. Springer, New York (2006)
4. Erradi, A., Anand, S., Kulkarni, N.: Evaluation of strategies for integrating legacy applications as services in a service oriented architecture. In: SCC 2006: Proceedings of the IEEE International Conference on Services Computing, Washington, DC, USA, pp. 257–260. IEEE Computer Society Press, Los Alamitos (2006)
5. W3C. Web services description language (wsdl) 1.1 (March 2001), `http://www.w3.org/TR/wsdl`
6. Dietz, J.L.G.: Enterprise Ontology, Theory and Methodology. Springer, Heidelberg (2006)
7. Ackermann, J., et al.: Standardized specification of business components. Memorandum of the working group 5.10.3, Component Oriented Business Application Systems (2002)
8. OASIS. Advancing web service discovery standard (June 2007), `http://www.uddi.org`
9. OASIS. Uddi products and components (June 2007), `http://www.uddi.org/solutions.html`
10. Colasuonno, F., Coppi, A., Stefano, Ragone, Scorcia, L., Di Noia, T., Di Sciascio, E.: juddi+: A semantic web services registry enabling semantic discovery and composition. In: The 8th IEEE Conference on E-Commerce Technology and the 3rd IEEE Conference on Enterprise Computing (2006)
11. Luo, J., Montrose, B., Kim, A., Khashnobish, A., Kang, M.: Adding owl-s support to the existing uddi infrastructure. In: ICWS 2006: Proceedings of the IEEE International Conference on Web Services (ICWS 2006), Washington, DC, USA, pp. 153–162. IEEE Computer Society Press, Los Alamitos (2006)
12. WSMO. D10 v0.1 wsmo registry (June 2007), `http://www.wsmo.org/2004/d10/v0.1/`

13. Rajasekaran, P., Miller, J.A., Verma, K., Sheth, A.P.: Enhancing web services description and discovery to facilitate composition. In: Cardoso, J., Sheth, A.P. (eds.) SWSWPC 2004. LNCS, vol. 3387, pp. 55–68. Springer, Heidelberg (2005)
14. Colgrave, J., Akkiraju, R., Goodwin, R.: External matching in uddi. In: ICWS 2004: Proceedings of the IEEE International Conference on Web Services (ICWS 2004), Washington, DC, USA, p. 226. IEEE Computer Society Press, Los Alamitos (2004)
15. Yan, Z., Cimpian, E., Zaremba, M., Mazzara, M.: Bpmo: Semantic business process modeling and wsmo extension. In: ICWS, pp. 1185–1186. IEEE Computer Society, Los Alamitos (2007)
16. Zaha, J.M., Albani, A.: Tool based support for teaching formal specification of business components. Teaching Formal Methods: Practice and Experience. Oxford, Great Britain (2003)
17. Fettke, P., Loos, P.: Specification of business components. In: Aksit, M., Mezini, M., Unland, R. (eds.) NODe 2002. LNCS, vol. 2591, pp. 62–75. Springer, Heidelberg (2003)
18. Baida, Z., Gordijn, J., Omelayenko, B.: A shared service terminology for online service provisioning. In: ICEC 2004: Proceedings of the 6th international conference on Electronic commerce, Delft, The Netherlands, pp. 1–10. ACM Press, New York (2004)
19. Active Endpoints Inc. et al.: Ws-bpel extension for people (bpel4people), version 1.0 (June 2007), `http://xml.coverpages.org/BPEL4People-V1-200706.pdf`

Appendix: Object Fact Diagram

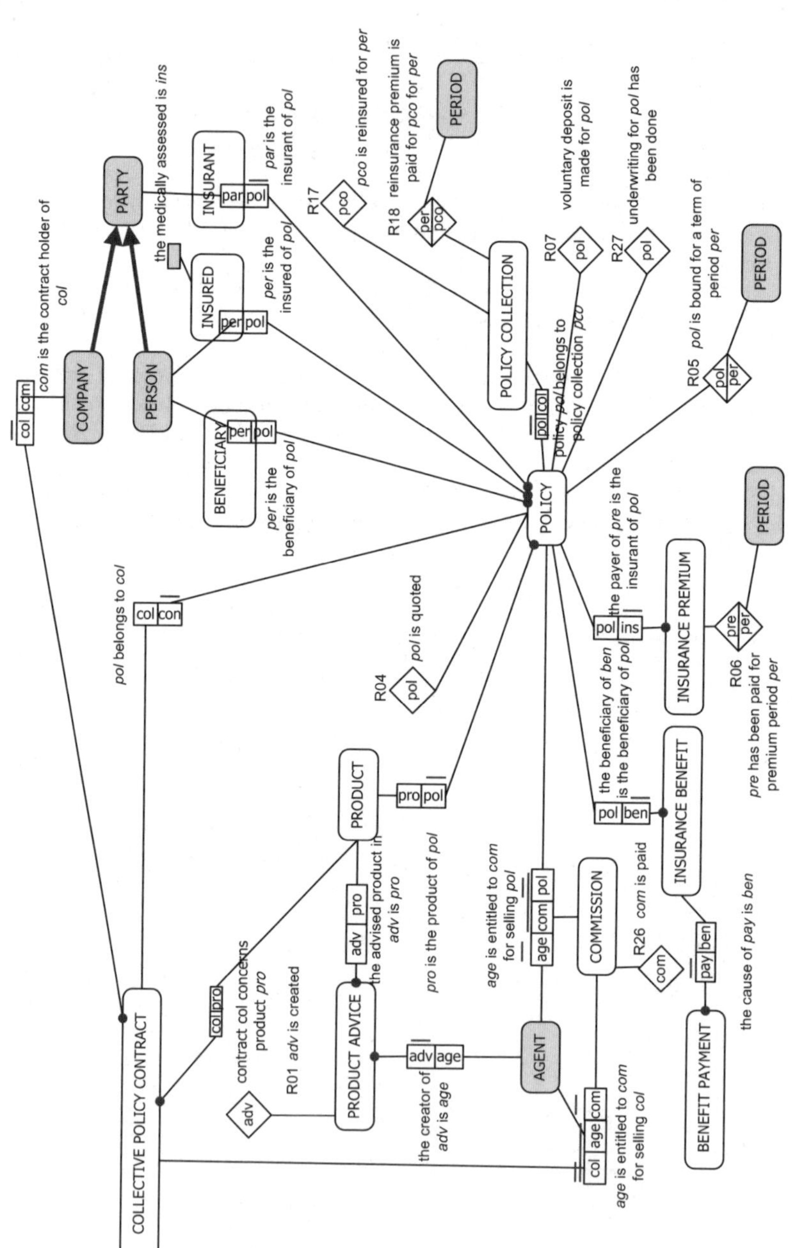

Fig. 3. Object Fact Diagram (OFD) of the life insurance company

Automated Model Transformations Using the C.C Language

Vojtěch Merunka[1], Oldřich Nouza[2], and Jiří Brožek[1]

[1] Czech University of Life Sciences Prague, Faculty of Economics and Management, Department of Information Engineering
merunka@pef.czu.cz, brozekj@pef.czu.cz

[2] Czech Technical University in Prague, Faculty of Nuclear Sciences and Physical Engineering, Department of Mathematics
nouza@km1.fjfi.cvut.cz

Abstract. This paper introduces the C.C programming language, which is designed for automated model transformations. The C.C language is outcome of our research and has been implemented in the CASE tool of a British software company. This technology started to be used by companies in the Central Europe and for business and software engineering courses at several Czech and Slovak universities, Loughborough University in the UK and Lehigh University in Pennsylvania. An interesting side-effect of this technology is C.C language application as a first teaching language in algorithmization, programming and software engineering courses.

Keywords: C.C language, model transformations, modelling and simulation, BORM.

1 Introduction

Modern CASE tools solve issues of business modeling and software modeling integration. This convergence requires strong support of the model-driven approach (MDA), where requirement modeling and business model simulation are used for subsequent information system design.

There are numerous modeling problems related to this matter. They concern interconnection of business models and software models, business process simulations, step-by-step transformations, domain-specific capabilities, flexibility, consistency and integrity checking etc. It remains questionable how these complex requirements should be implemented in CASE tools. We think that hard coding of these features is not the effective way. Hence we report our original experience with the C.C language, show its basic concepts, syntax and demonstrate the way it supports modelling process.

The C.C language design is an outcome of our research [11]. Interpreter of this language has been recently included into the Craft.CASE modeling tool developed by the British Company CRAFT.CASE Limited. This company thus takes all activities which were connected with the Craft.CASE and the BORM method in the past, including their future advancements.

J.L.G. Dietz et al. (Eds.): CIAO! 2008 and EOMAS 2008, LNBIP 10, pp. 137–151, 2008.

1.1 Model Transformation Techniques

There are several ways how to classify model transformation techniques. For example, Jean-Marc Jezequel [14] presents the following classification:

1. General purpose programming languages – Java, VB, C++, C#, etc. Rules and model behavior are implemented from scratch using the programming language.
2. Generic transformation tools – XSLT (XML transformation) and graph transformation tools.
3. CASE tools scripting languages – for example Arcstyler, Objecteering, OptimalJ, or Fujaba.
4. Dedicated model transformation tools – for example OMG QVT, which uses language OCL.
5. Meta Metamodeling tools – for example MetaEdit+, XMF-Mosaic, or KerMeta.

The Craft.CASE modeling tool provides model transformation via the C.C interpreter. This approach combines features from categories 1 to 4. In addition, the C.C interpreter is able to perform all operations on the model (including simulations, refactoring, new diagram creation, user-interactive procedures, manipulation with values of concrete object instances etc.), that are executed manually by users from graphical user interface. On the other hand, the language is not yet standardized on the present, therefore it is not possible to share the source code with other modeling tools.

2 Craft.CASE and BORM

Craft.CASE is a tool primarily targeted for modelling, testing and simulation of business processes and conceptual modelling of information systems using one coherent approach based in MDA and UML [3].

Craft.CASE and its C.C interpreter are developed in the VisualWorks for Smalltalk programming environment, which is an enterprise-class application development and delivery platform based on pure object-oriented programming [19].

The Craft.CASE implements the BORM method (Business Object Relation Modeling). BORM is the result of our previous work and has been described in [11,9,12]. BORM is based on the idea of object-oriented paradigm in conjunction with the process-based approach. As other MDA-based methodologies, BORM starts with a business-oriented specification of the problem area. Then it is step-by-step transformed to the correct software solution.

Craft.CASE supports these concepts transformations via business process simulators, instance-level modeling and set of transformation rules describing how to derive subsequent concepts from previous ones. Moreover, in each step of the method, Craft.CASE keeps consistency between two layers of a model; *subjects* and *behaviors*. Thanks to metamodel background and system internal C.C procedures, there is rigidly checked, whether all subjects from the first layer (e.g. classes, object states, packages etc.) have corresponding behaviors from the second layer (e.g. scenarios, use-cases, operations etc.) and vice versa.

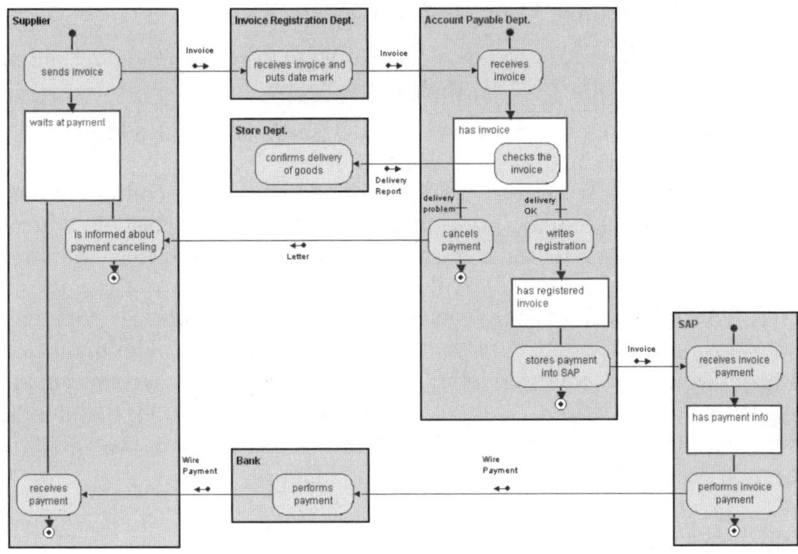

Fig. 1. Business process example

2.1 Business Process Modeling in Craft.CASE

Craft.CASE respects UML and MDA, but uses an original diagram for business process modeling. It conveys together information from three separate UML diagrams: state, communication and sequence. The BORM group have found that it is clearly understood by business stakeholders.

- Each subject participating in a process is displayed in its states and transitions.
- This diagram expresses all the possible process interactions between process partcipants. The business process itself consists of a sequence of particular communications and data flows among participating subjects.

More formally, BORM process diagrams are graphical representations of interconnected Mealy-type finite state machines [17] of particular subjects. Visual simulation of a business process is based on market-graph Petri net [18]. (This approach is described in [2].) Therefore we can show states, transitions and operations for all subjects playing a role in a business process. This is a very powerful, yet simple diagram.

There is an example of BORM business process diagram at figure 1. It shows an invoice processing. There are six participating subjects in this process: *Supplier*, *Invoice Registration Department*, *Store Department*, *Account Payable Department*, *SAP* and *Bank*. Small rectangles within these subjects are their states. Ovals within subjects are their activities, which are conceiveable as transitions between states as well. Big arrows between subjects are subject communications, small arrows are data flows.

2.2 Software Systems Modeling in Craft.CASE

Craft.CASE uses the UML standard for software systems modeling activities. The differences from other CASE tools are as follows:

1. Both business model and software model share common metamodel. This metamodel is described in section 4.1.
2. The content of UML software models is dependent on the content of business models. For example, each conceptual class from UML class diagram must have its predecessor in some business subject. This information is kept in the project database.
3. Craft.CASE stresses pre-implementation stages of system development and is not primarily appointed for software code generation. The main purpose is business and software modeling. This is the reason for strong demand on consistency, simulation, cross-reference checking and other features. However, various source codes generated from the project database are feasible through user-written modules in the C.C language.

3 The C.C Language

The C.C language is a functional programming language with PASCAL-like syntax with several imperative constructs and some features coming from languages PROLOG, Erlang, Ruby, Python and Smalltalk. It has an interpreted programming environment. C.C is used for following purposes:

- As a scripting language. Procedures in C.C are able to pass through project database and compose miscellaneous documentation reports.
- Precise process simulation. Procedures in C.C can compute various simulation data, control simulation flow, etc.
- Automated manipulations with the model (e.g. applying design patterns, refactoring and class normalization).
- Consistency and integrity check of project database. This feature covers the same functionality as the OCL [4].
- Data export in different formats (namely XMI and binary formats of other CASE tools).
- Data import from different data sources (e.g. ODBC, CSV etc.).

3.1 Inspiration – Pascal and LISP

The Pascal programming language as a simplified ALGOL was so successful that there were and still are attempts to use its ideas in many application areas. These include miscellaneous languages for data manipulations, scripting or object-oriented programming. Pascal is still frequently used as a first programming language. Its syntax dominates in theoretical publications oriented at algorithmization, theory of programming, new algorithm description etc.

The LISP programming language has been developed out of the needs of artificial intelligence programming. LISP (and its modern successor Scheme) smoothly implemented the concepts of lambda-calculus and functional programming introduced by Alonzo Church in 1930s [8].

3.2 Motivation

As stated in section 1.1, the C.C language covers features of miscellaneous existing tools. We required these features to have in one modeling environment. First of all, these are

- interactive model transformations,
- prototyping and instance-level testing (e.g. querying and manipulation with concrete values of particular object instances emerging in the model) and
- simulation.

Recently we have decided to use C.C as the same language for algorithmization and software engineering courses.

3.3 Basic Concepts

- The C.C architecture consists of *modules* having *functions*. Modules are both system built-in and user-written.
- Variables must begin with capital letters.
- Built-in values are `true`, `false`, `nil`, `e` (Euler's number), `i` (purely imaginary number), `pi` (Ludolf's number), `infinity`, `tiny` (infinitesimal zero) and a lot of functions in miscellaneous modules.
- The only types are:
  - Symbol (atomic textual values beginning with non-capital letters).
  - String of characters written in double quotation marks.
  - Number.
  - Date in format `DD-MMM-YYYY`.
  - Time in format `HH:MM:SS`.
  - Logical value as predefined symbols `true` and `false`.
  - `nil`.
  - Collection of elements. There are three types of collections: `list`, `set` and `dictionary`.
  - Function, which implements lambda-expression and is written in curly brackets. To illustrate, lambda-expression $\left(\lambda x \lambda y \mid x^2 + y\right)$ is written as `{:X,:Y|X^2+Y}`.

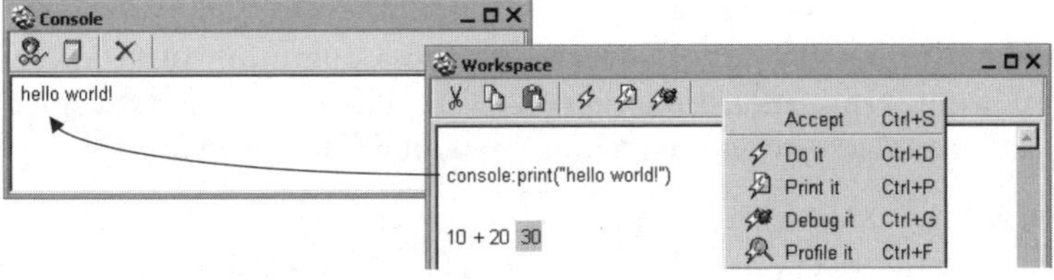

Fig. 2. Hello world example in C.C

Following line of the C.C code implements a hello world program (shown in the figure 2). This code is written in the system *workspace* and the result appears in the system *console*.

```
console:print("Hello world!").
```

3.4 Functions

Each function must be a member of a certain module. Therefore the previous example of a hello world program working with the module named `console` and the function named `print` can be written as follows.

```
| M , F |                # declaration of variable names
M := console.            # assigning symbol "console" to var "M"
F := print.              # assigning symbol "print" to var "F"
M:F(hello world!).   # function call
```

User-defined functions are represented by function expressions stored in variables. For example function $F(x, y) = 10x + y$ can be implemented as

```
F := {:X , :Y | 10*X + Y}.
```

The `F` function can be applied on arguments via round brackets as for example `F(3,4)`. However, this function call can be used directly without the need to store this function in any variable, like `{:X,:Y|10*X+Y}(3,4)`.

(Additionally, there are yet some advanced features related to default values of lambda-variables, order of parameters in function call and possibility to call function with incomplete set of parameters).

3.5 Collections

Following example shows declaration of a list `L` and a dictionary `D`.

```
L := [10 , 20 , 30 , 40 , 50].
D := [first := 10 , second := 20].
```

Then we can access elements of these collections as follows.

```
L[1] = 10.   L[2] = 20.   D[first] = 10.   D[second] = 20.
```

We have defined nine operators for comfortable collection processing (e.g. element adding, removing etc.). Nonetheless these operators (and all other C.C operators) are interpreted as functions as well. There are also two special functions *selection* and *projection*, which are explained in this example:

```
[10,20,30,40,50] // {:X | X > 20}   =   [30,40,50].
[10,20,30,40,50] >> {:X | X + 1}    =   [11,21,31,41,51].
```

Craft.CASE model elements behave as collections as well. For example, if there is an element `AClass`, then the expression `AClass[name]:=NewValue.` changes the value of the property `name` of this element.

3.6 Control Structures

Control structures are realized by operators, but they have internally the same interpretation as functions. They are:

- `if` *logical-expression* `then` *function* [`else` *function*]`.`
- `for` *collection* `do` *function*`.`
- `from` *value* `to` *value* [`by` *value*] `do` *function*`.`
- `repeat` *function* `until` *logical-expression*`.`
- `while` *logical-expression* `do` *function*`.`

Following two pieces of code show the same iteration:

```
for [10,20,30,40,50] do {:X | console:print-nl(X)}.

| X |
X := 10.
while  not(X > 50)  do  {console:print-nl(X). X := X + 10}.
```

3.7 Programming Environment, C.C Data Modeler

Programming environment of the C.C implementation consists of *module browser* and *traceable debugger* (figure 3) and *time profiler* (figure 4).

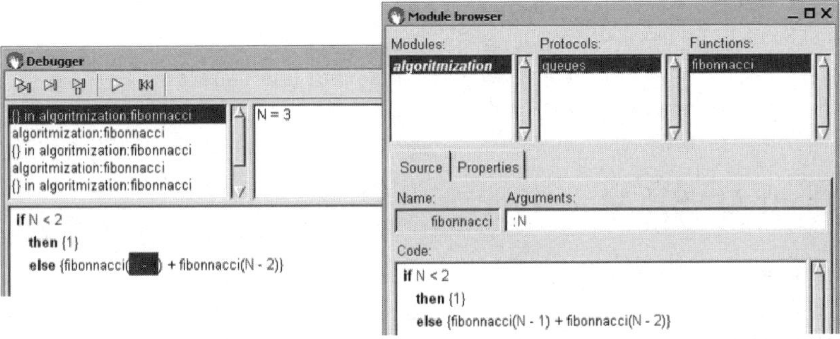

Fig. 3. Debugger and module browser

The tool also has a lite-version called C.C Data Modeler supporting the C.C language and only selected subset of UML models. This version is aimed for software engineering and introductory programming courses.

3.8 Code Examples

The C.C language is a universal programming language. It is used at several universities as an introductory programming language. Here are two small examples of well-known algorithms.

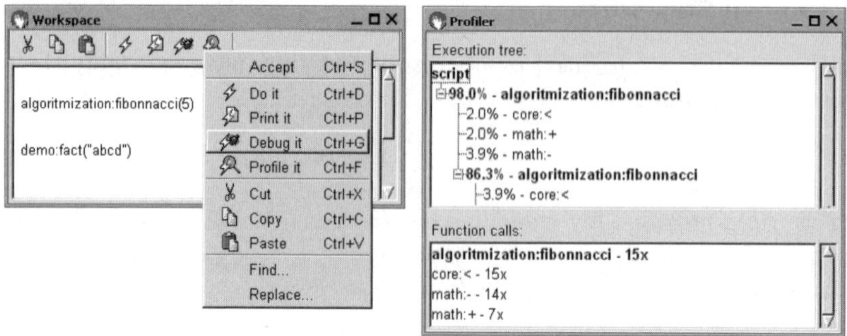

Fig. 4. Profiler

```
##### recursive definition of factorial        #####
| Factorial |
Factorial := {:X |  if X = 0
                  then {1} else {X * Factorial(X - 1)}}.

##### Eratosthenes' generator of prime numbers #####
| Max , Non-primes , Primes |
Max := integer(dialog:request("maximum number?")).
Non-primes := set:new().   Primes := list:new().
from 2 to Max do {:N |  if not(N in Non-primes)
                      then {Primes add N.
                            from N to Max by N
                              do {:N1 | Non-primes add N1}}}.
console:print(Primes).
```

4 Craft.CASE Modeling

4.1 Craft.CASE Metamodel

Craft.CASE works with a simple metamodel, which is common for both business and software modeling. This model shares the same ideas as other metamodels used in CASE tools implementations. (For example GOPRR metamodel [10] by Steven Kelly made for MetaEdit+ CASE tool [13]).

In the Craft.CASE metamodel we work with only two types of *elements*: *nodes* and *links*. Each *link* is one-way oriented and has one *source* and one *target*. Content of *source* or *target* can be both *node* or *link* (See figure 5). Each type of *link* knows which are its possible types of *sources* and *targets* to be connected.

The whole project is a *node* as well. If this project consists of diagrams, they are *nodes* linked to this project. If a diagram consist of elements, they are *nodes* linked to this diagram. Of course, miscellaneous relations between the elements in particular diagrams are *links* between corresponding *nodes* too.

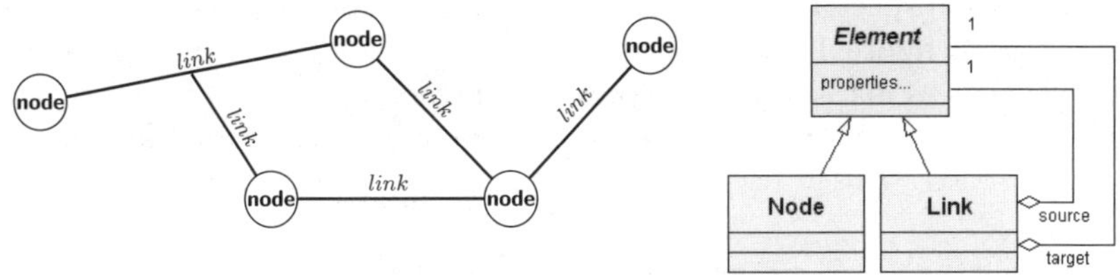

Fig. 5. Craft.CASE metamodel

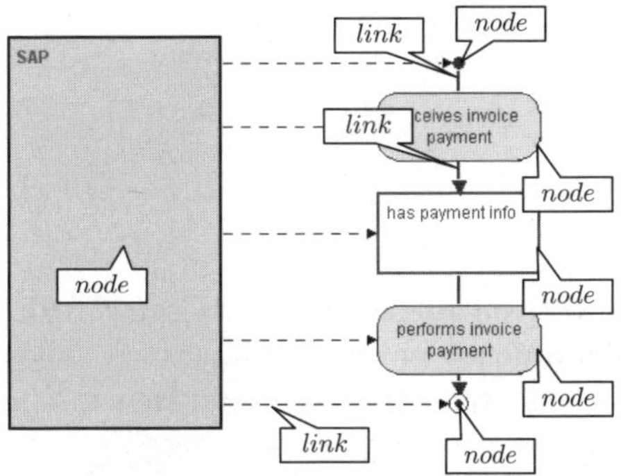

Fig. 6. Concrete nodes and links

Figure 6 shows the Craft.CASE metamodel on the example of business process from figure 1. Subject *SAP* is a *node* connected to other five *nodes*. These five *nodes* are states and activities of this subject *SAP* and are connected by *links*.

Another example of the metamodel can be the UML class-diagram. Class symbols are *nodes*, relations (e.g. inheritance, association, composition, etc.) are *links* between classes. If a class has methods (or attributes, of course), these methods are *nodes* connected to this class. Additional information like cardinalities or link attributes are *nodes* connected to corresponding association, which is a *link* as well, because it connects classes.

4.2 C.C Constructs for Craft.CASE Metamodel

The Craft.CASE metamodel is supported by several system modules. There is also one special operator (internal function) called "path expression" in the metamodel structure. This operator complements functionality of selection and projection. Semantics of selection operator is also extended.

- *element-or-collection-of-elements* `->` *function-or-typename*.
- *collection-of-elements* `//` *function-or-typename*.

Path expression is an implementation of the graph traversing algorithm. It contrives to collect neighbors of an element or a collection of elements in the project database with respect to the metamodel. For example, if there is an element `Supplier` from figure 1, then we are able to collect all interrelated data-flows of this element. This is a path expression traversing from this element to its activities, then from these activities to their communications and finally to their data-flows.

```
Supplier -> "activity" -> "communication" -> "data flow"
                       =  ["Invoice" , "Letter" , "Wire Payment"].
```

Another example shows the selection of only the data flow elements from a set of all selected elements in a diagram editor.

```
editor:selection()  //  "data flow".
```

4.3 C.C Modules

The interpreter of the C.C language made for Craft.CASE has several built-in modules with a certain amount of built-in functions. These functions are implemented directly in the system byte-code [19]. Hence their execution is much faster. These functions cover general programming features (this segment of C.C can be used separately from Craft.CASE), support for precise simulations and support for the Craft.CASE metamodel and user interface (e.g. dialogues, console output, file access, ODBC data access etc.). There are yet some interesting modules:

- `math`: Support for mathematical processing such as work with complex numbers, vectors, matrices, infinitezimal numbers, numerical derivation and numerical integration etc. For example, `math:integral({:X|X^2+X},3,5)` denotes $\int_3^5 (x^2 + x)\,dx$.
- `list`: Functions specific to lists of values as `head`, `tail`, `cons`, `sort`, etc.
- `project`: Basic processing with the project database. Here are functions as `new-node`, `new-link`, `remove-element`, for example.
- `element`: Functions specific to project elements. Here are functions as `nodes`, `links`, `source`, `target`, `get-property`, `set-property`, for example.
- `diagram`: Functions specific to diagram elements. For example, here are functions as `add-element` and `remove-elements`. (If an element is removed from a diagram, still it can persist in the project database.)
- `editor`: Functions specific to an actively opened diagram editor. Here are functions as `selection`, `add-into-selection`, `remove-from-selection`, etc.
- `simulation`: Functions specific to business process simulation. Here are functions as `activate`, `start`, `step`, `terminate`, `raise-exception` etc.
- `report`: Functions for generation of Craft.CASE reports in formats HTML and PDF.

5 Modeling Examples

Our experience denotes the fact that the design pattern technique, the object normalization technique and refactoring technique share the common principle of model transformation. Hence all these techniques can be automated through the C.C code with a project database. In this chapter we demonstrate practical examples of this idea. These examples are taken from our software engineering courses.

5.1 Refactoring

It is possible to define refactoring as any sequence of system transformations, where the behavior of the system remains unchanged. (An exception might be for instance a slightly different delay between the user impulse and the subsequent system response, nevertheless from the user's point of view, refactoring has actually no importance.) From the system modeling aspect, refactoring is performed for optimalization, reusability and maintainability reasons [15]. A classical book on refactoring is [5].

In the following piece of code we present an interactive algorithm for creating a new super-class to selected classes from a conceptual class-diagram.

```
| Classes , NewClass |
# which are classes from selection?
Classes := editor:selection() // "Class".
if list:is-empty(Classes)then{return dialog:warn("No classes selected!")}.

# create new class, name it and add it into diagram
NewClass := project:new-node("Class").
NewClass[name] := dialog:request("New class name?").
editor:add-element(NewClass).

# assign new class as superclass of selected classes
for Classes
 do {:Class | editor:add-new-link(Class , NewClass , "Supertype")}.
```

5.2 Design Patterns

Design patterns are proven solutions to design problems. A design pattern is a template for how to solve a particular problem. It is mature, proven and widely accepted software development technique. More information on design patterns for software systems development and their classification is in the book [6].

Currently all design patterns from this book are implemented as interactive functions in the C.C language. Moreover, we expect existence of similar patterns for business process modelling. This is indicated in [2], for example. Hence we started the exploration of business process patterns from practical projects made in the Craft.CASE and their subsequent implementation as C.C functions.

Following piece of code shows the implementation of *Adapter* pattern, which is demonstrated in pictures 7 and 8.

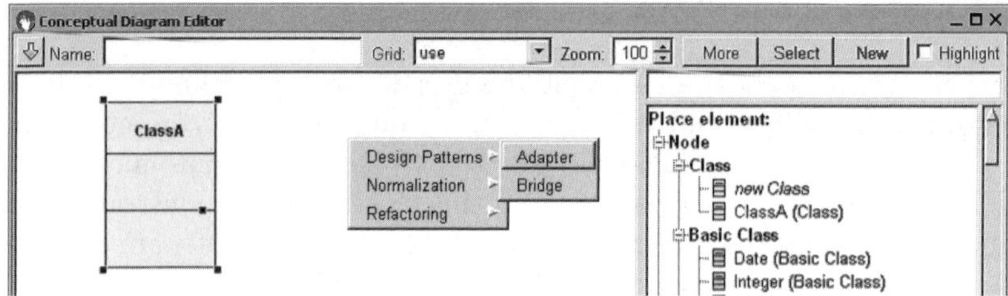

Fig. 7. Adapter pattern example - initial situation

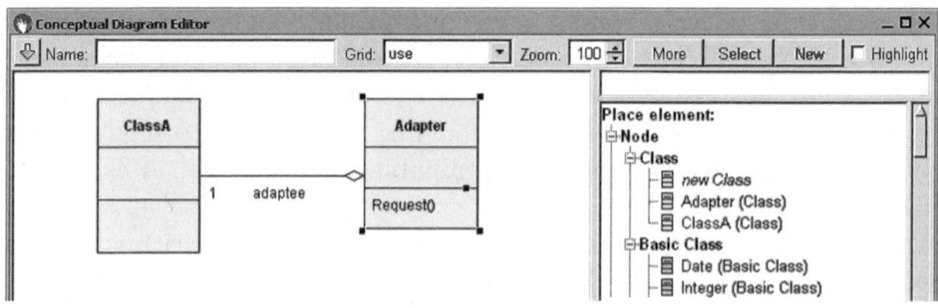

Fig. 8. Adapter pattern example - result

```
| Classes , Adaptee , Adapter , AdapterLink , Method |
# which are classes from selection?
Classes := editor:selection() // "Class".
  if list:size(Classes) <> 1
then {return dialog:warn("Select one class to be adapted!")}.

# select adaptee class
Adaptee := Classes[1].

#create and link adapter class
Adapter := project:new-node("Class"). Adapter[name] := "Adapter".
editor:add-element(Adapter).
AdapterLink := project:new-link(Adapter, Adaptee , "Composition").
AdapterLink[name] := "adaptee". AdapterLink[cardinality] := "1".
editor:add-element(AdapterLink).
Method := project:new-node("Method"). Method[name] := "Request()".
project:new-link(Adapter , Method,  "ownership (conceptual)").
```

5.3 Object Normalization

Object normalization is a similar approach to the relational data normalization. It is applied to object-oriented data model. There are several approaches to object normalization [16]. The most advanced information is in the book by Scott Ambler [1], where object-oriented data modeling is discussed. In our technology,

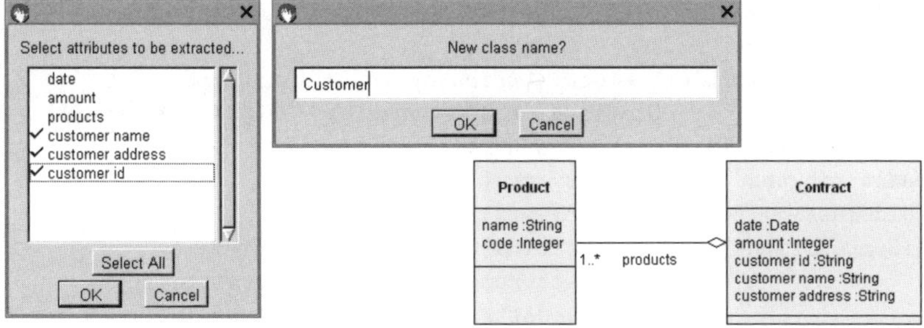

Fig. 9. Third normal form example - initial situation

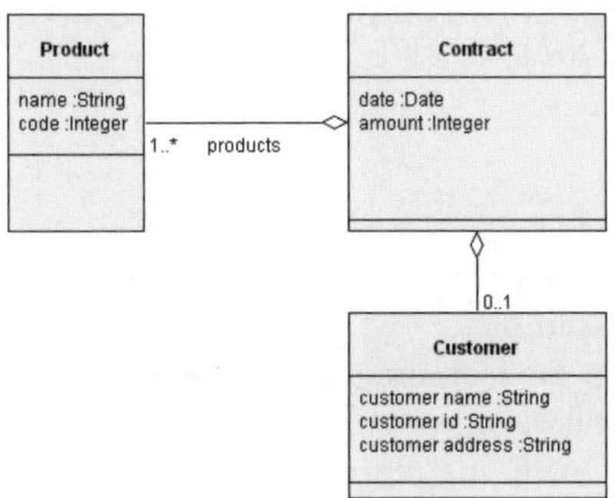

Fig. 10. Third normal form example - result

we use the Scott Ambler's three levels of object-oriented model normal forms. (It is obvious that "standard" relational normal forms are implementable by a C.C code as well.)

Following user-interactive function shows our implementation of the third normal form. Figures 9 and 10 show a concrete example of extracting new class `Customer` from class `Contract`.

```
| Classes,OldClass,NewClass,AttributeNames,RemovedAttributes,Link |

# which is class in selection?
Classes := editor:selection() // "Class".
  if list:size(Classes) <> 0
then {return dialog:warn("Select one class!")}.
OldClass := Classes[1].

# create new class, name it and add it into diagram
NewClass := editor:add-new-node("Class").
NewClass[name] := dialog:request("New class name?" , "New Class").
editor:add-new-link(OldClass , NewClass , "Composition").
```

```
# select instance variables to be extracted from old class to new class
AttributeNames :=
dialog:choose-multiple("Select attributes to be extracted...",
                       OldClass -> "Composition" >> {:X | X[name]}).
#remove them from an old class and remember them
RemovedAttributes := dictionary:new().
for (OldClass -> "Composition")
 do {:Composition |  if Composition[name] in AttributeNames
                   then {project:remove-element(Composition).
                         RemovedAttributes[Composition[name]] :=
                                       element:target(Composition)}}.
#link removed attributes into a new class
for dictionary:keys(RemovedAttributes)
 do {:Name | Link :=
        project:new-link(NewClass,RemovedAttributes[Name],"Composition").
        Link[name] := Name}.
```

6 Conclusion

In this paper, we presented the C.C language concepts and demonstrated the C.C modeling on real examples.

The C.C language is an instrument we use to support our research in the area of business and software systems modeling. The combination of the language C.C, the Craft.CASE metamodel and BORM methodology makes us a flexible technology, which allows to model business and software systems in one coherent paradigm. It solves the interconnection of business models and software models, business process simulations, step-by-step model transformations, domain-specific capabilities, model checking an reporting etc.

Our next work will concentrate on elaboration of the C.C language, connectivity to other tools and research in the area of business process patterns.

Acknowledgement. The authors would like to acknowledge the support of the Czech Ministry of Education, Youth and Sports by the grant project MSM6046070904 and LA08015.

References

1. Ambler, S.: Building Object Applications That Work, Your Step-By-Step Handbook for Developing Robust Systems Using Object Technology. Cambridge University Press/SIGS Books (1997)
2. Barjis, J.: Developing Executable Models of Business Systems. In: Proceedings of the ICEIS - International Conference on Enterprise Information Systems, pp. 5–13. INSTICC Press (2007)
3. Craft.CASE home page, http://www.craftcase.com
4. Pollet, D., Vojtisek, D., Jezequel, J.-M.: OCL as a core UML transformation language. WITUML 2002 Position paper, Malaga, Spain (2002), http://ctp.di.fct.unl.pt/ja/wituml02.htm

5. Fowler, M.: Refactoring. Addison-Wesley, Reading (1999)
6. Gamma, E., Helm, R., Johnson, R., Vlissides, J.M.: Design Patterns - Elements of Reusable Object-Oriented Software. Addison-Wesley, Reading (1994)
7. Hall, J., et al.: Accounting information systems - Part 4, System development activities, 4th edn., Thomson South-Western New York (2004)
8. Hankin, C.: Lambda Calculi - A Guide for Computer Scientists. Clarendon Press, Oxford (1994)
9. Liu, L., Roussev, B., et al.: Management of the Object-Oriented Development Process - Part 15: BORM Methodology. Idea Group Publishing (2006)
10. Kelly, S.: GOPRR Metamodel - Appendix of Towards a Comprehensive MetaCASE and CAME Environment: Conceptual, Architectural, Functional and Usability Advances in Metaedit+. Ph.D. Thesis. Jyvskyl University (1997)
11. Knott, R.P., Merunka, V., Polak, J.: The BORM methodology: a third-generation fully object-oriented methodology. Knowledge-Based Systems 16(2), 77–89 (2003)
12. Merunka, V., Polak, J., Knott, R.P.: Process Modeling for Object-Oriented Analysis Using BORM Behavioral Analysis. In: Proceedings of Fourth International Conference on Requirement Engineering - ICRE 2000, IEEE Computer Society, Los Alamitos (2000)
13. MetaEdit+ home page, `http://www.metacase.com/`
14. Muller, P.-A., Fleurey, F., Jezequel, J.-M.: Weaving executability into object-oriented meta-languages. In: Briand, L.C., Williams, C. (eds.) MoDELS 2005. LNCS, vol. 3713, pp. 264–278. Springer, Heidelberg (2005)
15. Sunye, G., Pollet, D., Le Traon, Y., Jezequel, J.-M.: Refactoring UML Models, `http://www.irisa.fr/triskell/publis/2001/Sunye01b.pdf`
16. Vrany, J., Struska, Z., Merunka, V.: Object normalization as the contribution to the area of formal methods of object-oriented database design. In: Proceedings of the eighth International Conference on Enterprise Information Systems: Databases and Information Systems Integration ICEIS 2006, Paphos, Cyprus, INSTICC Press (2006)
17. Roth Jr., C.H.: Fundamentals of Logic Design. Thomson-Engineering, 364–367 (2004)
18. Peterson, J.L.: Petri Net Theory and the Modeling of Systems. Prentice-Hall, Englewood Cliffs (1981)
19. VisualWorks home page, `http://www.cincom.com/visualworks/`

Improvement in the Translation Process from Natural Language to System Dynamics Models

Yutaka Takahashi

School of Commerce, Senshu University
2-1-1, Higashimita, Tama, Kawasaki, Kanagawa, Japan
takahasi@isc.senshu-u.ac.jp

Abstract. In order to solve social science problems or make models for business forecasts, descriptive information in natural language is as important as measured numerical information. While numerical information is widely used in various stages, most descriptive information is employed only to describe the background or the very primitive stage of modelling. However, descriptive information has rich contents. One uses it to organise one's ideas and to communicate with other people. The modelling method which can employ this rich content information is System Dynamics. It has an interface to express such descriptive information, stock flow diagrams. Moreover, these diagrams are also used when numerical simulations are conducted. However, matching descriptive information and diagrams has depended on model builders' skill and mental models. This paper shows matching rules between descriptive information and stock flow diagrams as improvement in an existing method.

Keywords: Descriptive information, natural language, simulation.

1 Introduction

Research themes of social science contain numerical and descriptive information. Both kinds of information are equally important in building models which are employed to make solutions or finding the nature of social systems. Numerical information is usually processed by techniques similar to ones in natural science fields. However, descriptive information is mainly used only in the earlier stages of modelling, such as setting the direction of research.

Nevertheless, descriptive information naturally has rich contents, which are essential for meaningful research; most descriptive information expresses fundamental value or mental models, which are important to make formal model precisely, of involved people. This paper shows some guidelines to find dynamic structure and mathematical information from descriptive information.

Research objects of natural science have enormous numerical data so that studies can be completed without descriptive information. Moreover, natural science fields have no "interactive" research objects; scientists need to watch and record their research objects but do not need to interview animals, plants, or inorganic materials.

J.L.G. Dietz et al. (Eds.): CIAO! 2008 and EOMAS 2008, LNBIP 10, pp. 152–163, 2008.

Therefore, it is reasonable that there is no need to process descriptive information in natural science.

On the other hand, social science fields have enormous amounts of descriptive information, in addition to numerical information. For example, fieldwork researchers conduct interviews. Interviewees' stories are mainly descriptive. In business fields, company managers have their own opinions, beliefs, or mental models[1]. These kinds of information are all descriptive information. Researchers should be attentive to them in order to organise their opinions. Since what social science researchers are studying tends to have more hidden or complex causal structures than research objects of natural science, the role of descriptive information is important in order to obtain appropriate analyses.

However, descriptive information is not easily processed in the same way as numerical data. First, descriptive information is written in natural language, which is used when one records something or communicates with others, such as English, French, etc. Natural language is flexible both in terms of reading and writing, so some possibilities for misunderstanding exist. Second, the interpretation or evaluation of something, expressed in stories, depends on the reader.

Such difficulties are recently found not only in social science fields but also nearby natural science fields. For example, software development processes are logical and scientific, but software must match users' needs. Therefore, there are several practices which use requirement analysis methods, such as UML [1]. These methods or tools focus on leading software development projects. This means that communication (or whole information flows) is relatively close and takes only a short time; therefore, hidden effects caused by feedbacks are not considered. This ignorance of feedback effects does not cause problems when managing a short software development process.

On the other hand, social science fields, including business activities, policy making and social movement analyses, need to consider long term effects of past activities or decisions and mutual effects not only inside but also outside object systems.

A method which can manage this problem is System Dynamics modelling. System Dynamics is described as "a rigorous method for qualitative description, exploration and analysis of complex systems in terms of their processes, information, organisational boundaries and strategies," by Wolstenholme [2]. Models are expressed as stock flow diagrams which indicate causal structures and types of variables. Stock flow diagrams express how model builders understand their research objects and produced models are constructed. Fig. 1 is one of the simplest examples.

It indicates that population increases only when babies are born and the number of babies is influenced by the current population and capacity. There is no other interpretation. The model must reflect its model builder's idea about the object, namely a population system. Thus, stock flow diagrams can trace descriptive information and avoid misunderstandings brought about by the ambiguity of natural language.

---

[1] Mental models are firm beliefs or stereotypes which are not derived from logical deduction, similar to ideas expressed by Senge [4].

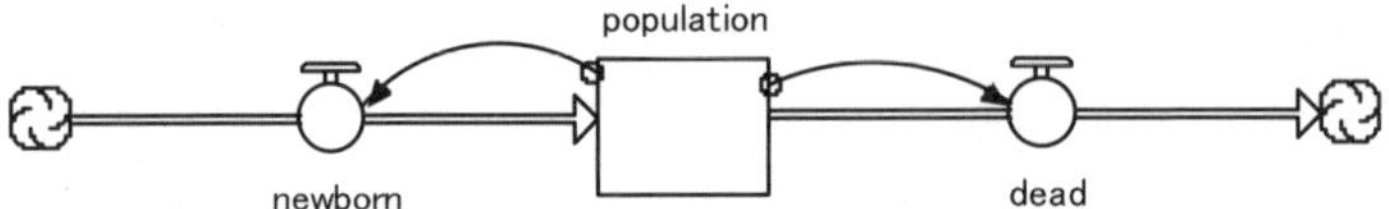

Fig. 1. Here indicates an example of a stock flow diagram, based on a story about population dynamics. Of course, this is not a real, practical model, but includes all fundamental elements in System Dynamics models.

There is another diagram in System Dynamics: namely causal loop diagrams. Causal loop diagrams indicate only causal structures in models. Fig. 2 is a causal loop diagram sample using the same theme as fig. 1. One can conduct qualitative analyses using causal loop diagrams. Causal loop diagrams are easily drawn even by beginners of System Dynamics. In addition, qualitative analyses can allow researchers to have an open mind and present various perspectives of elements in their research. However, most practical results require quantitative simulation, which causal loop diagrams cannot provide but which stock flow diagrams can, because parameter volumes influence performance of systems. Therefore, it is ultimately necessary to make stock flow diagrams and to perform some simulations in order to obtain meaningful results.

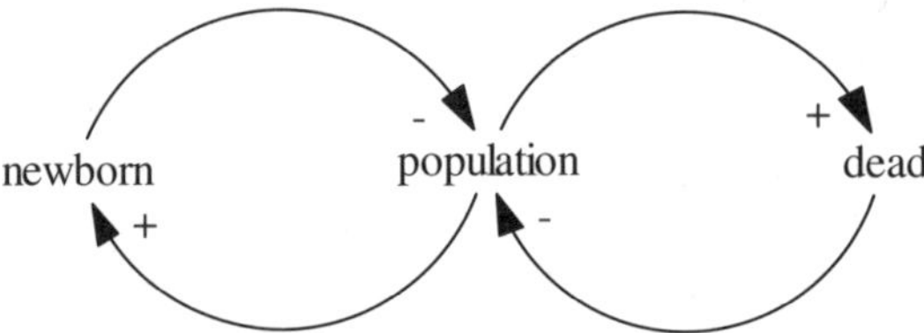

Fig. 2. The causal loop diagram's structure is very similar to that of the stock flow diagram. However, an arrow which decreases population has the opposite direction. Stock flow diagram's pipes, or double line arrows mean flow-in and flow-out. However, all arrows in causal loop diagrams express the cause-effect (or dependent-independent) relationships.

Stock flow diagrams are similar to causal loop diagrams. Nevertheless, some people who can draw causal loop diagrams face difficulties in drawing stock flow diagrams. The difference between stock flow diagrams and causal loop diagrams is the presence of the variable type classifying process. Therefore, one reason for the difficulty in modelling stock flow diagrams is the difficulty in distinguishing types of variables: stock, flow or auxiliary. Indeed, Sweeney and Sterman [5] reports that people are not used to dealing separately with stock and other variables.

A classic method to distinguish types of variables is the "snapshot test," explained in Sterman [3]. This test makes logical sense. However, there are sometimes possibilities that a variable can be defined both as a stock and as another type. It depends on model builders' understanding of objects. Thus, this approach is directly influenced by the ambiguity of natural language.

As another approach, Richmond [6] indicates correspondences between elements in natural language and types of variables in stock flow diagrams. He suggestes the metaphors that stock, flow and auxiliary variables are namely subjects, verbs and adjectives in natural language sentences. These are not rigid correspondences but heuristically convenient for model builders. The use of descriptive information as important parts of models or modelling is unique. UML modelling texts, such as Jacobson et al. [1], also mention similar approaches[2].

These correspondences contribute to easy distinction when deciding types of variables. This idea is instructive, but does not mention relationships among variables. Takahashi [7] suggests an extension of Richmond's idea. He points out sentence patterns corresponding to eight combinations of variables in stock flow diagrams. Possible combinations in stock flow diagrams are limited to these eight patterns. Each corresponding sentence is defined using simple syntax and limited vocabulary. Model builders paraphrase their descriptive information using these simple syntax sentences; simple syntax sentences are used as a middle language. Both original descriptive information and simple syntax sentences are written in natural language. Model builders can paraphrase without special tools.

However, there is a possibility that multiple sentences can apply to one combination. For example, the sentence "More A, more B" can produce any combination of variables. It is possible to avoid this problem by applying more strict (narrow meaning) sentences earlier than sentences with wide meaning. However, when model builders assign sentences with wide meaning, it can be difficult to change them after finishing paraphrasing.

This study shows rules to use correspondences appropriately in Takahashi [7] and suggests an improvement on them.

2 Rules to Select an Appropriate Stock Flow Structure

The fact that one sentence has multiple choices of model structure can lower its convenience in using the correspondences suggested in Takahashi [7]. The reason that this problem arises is a lack of two important concepts in the middle language: units, delay, and expression of transition of an individual material or a person.

2.1 Unit Concept

System Dynamics modelling needs to deal precisely with units of all variables, in the same way as other modelling methods do. This means that a stock variable and its flow variables must use an identical unit. This consistency of unit keeps model builders from excessive simplification. For example, suppose we are given the descriptive information below.

"More employees, more output."

This sentence contains two variables: employees and output. If this sentence expresses a scene in a steel factory, these two variables obviously have different units; employees should be counted as a number of people and output might be tons. Therefore, model builders reasonably divide them into two separate stock flow combinations (Fig. 3).

---

[2] They suggest making classes by finding nouns in requirement specification documents in Object Oriented Programming.

Richardson and Pugh [8] mentions that this check can "prove incorrectness of wrong modelling." In the meaning that inconsistent use of units in a stock flow combination, this check works correctly. However, if all variables deliberately use the same unit, the check does not work. For example, see the sentence below.

"More order placement, more shipment."

One can use the same unit for both variables: order and shipment. Indeed, to use dollars or a number of goods is reasonable for most cases for the scenario expressed by this sentence. However, it is inappropriate to put both variables in the same stock flow combination. Fig. 4 is an incorrect diagram for the sentence above.

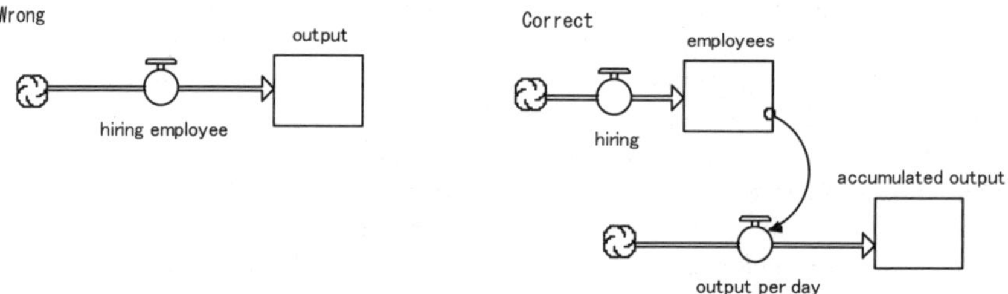

Fig. 3. On the left is an incorrect stock flow diagram. It should be drawn in separate stock flow lines.

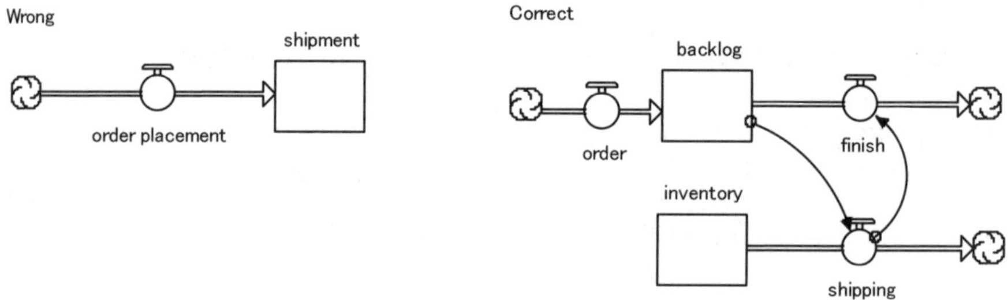

Fig. 4. The number of Order placement is not the amount of sales goods

Moreover, this diagram appears to mean and actually means "order placement" activity brings sales goods to the shop shelf. This is strange and actually different from reality.

Indeed, they have the same unit. However, the number of order placement is information, while the amount of shipment is the number of sales goods. This distinction between information and physical quantity and understanding what kind of material flows in it is significant in avoiding this problem. Richmond [6] suggests a "unit conversion flow." This idea simply brings the difficulty of correct modelling to model builders because it eliminates unit check possibilities.

Therefore, in addition to the unit consistency rule, model builders need to deal separately with information variables and physical quantity variables. In addition, each stock flow combination carries and reserves only the same material in it when a translation using simple syntax middle language is conducted.

2.2 Delay and Causality

Causal relationships, which start from stock variables, implicitly suppose time delay. This is natural and reasonable because of the meaning of causality; all results follow their causes. In this sense, relationships without any time delay are not causal relationships but an "expression in other words[3]".

Most System Dynamics software has "delay" functions. These functions are originally "abbreviations" of some stock flow structures which have causal relationships starting from stock variables.

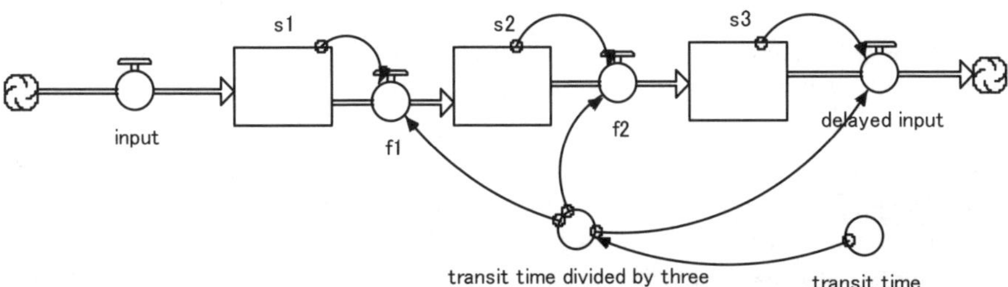

Fig. 5. "The third order exponential delay" can be expressed without built-in functions

Moreover, relationships from outflow variables to other variables also have implicit stock flow structures. Some software can set outflow variables as "non negative." This means that these outflow variables refer the state of connected stock variables. If descriptive information refers to these stock variables' change, they therefore implicitly include time delay.

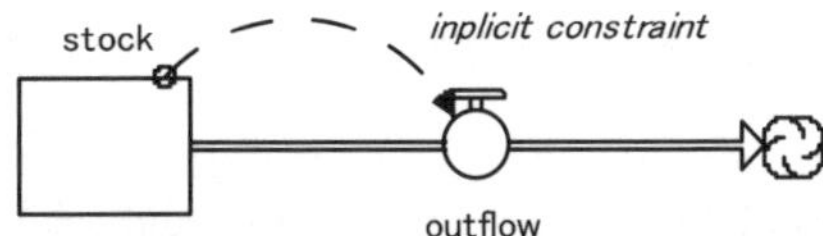

Fig. 6. Outflow variables are controlled by their stock variables when they are set as "non-negative," even if there is no explicit link indicated as the broken line arrow

Thus, when descriptive information can be correctly expressed with delay, there should be stock variables in corresponding models.

2.3 Transition of Individual Material or Person

System Dynamics modelling allows model builders to make sequences of stock flow connections. Direct use of the method suggested by Takahashi [7] produces only separate

[3] Some software such as STELLA and ithink (isee systems) uses the variable type name "converter" instead of "auxiliary."

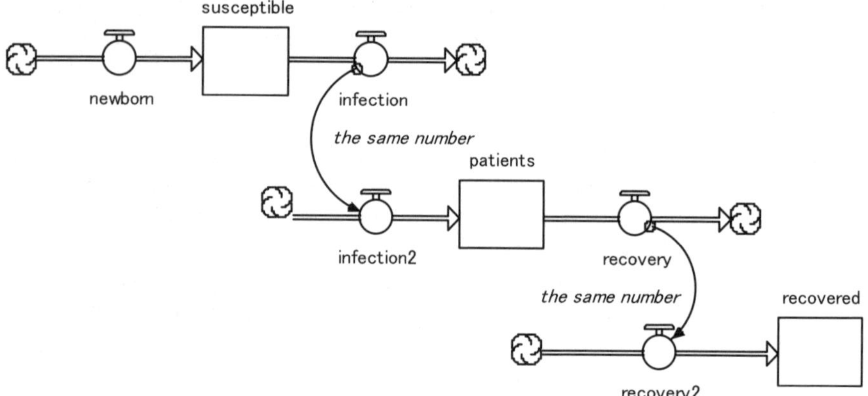

Fig. 7. Each stock indicates the population in a stage of immunising. Some separated stock flow combinations should be connected as a sequential pipeline.

"one stock with some flow variables structures" with reference links, not with flow links such as in Fig. 7. It should be connected for the reason of simplicity.

To determine the connection of stock flow combinations, each stock variable in these combinations must store the same elements at different time. If an individual material or person can be counted in a unit all time and transit several states over time, each state should be expressed as a stock variable, and transition movements between stocks should be expressed as flow variables. For example, Fig. 7 should be expressed as in Fig. 8.

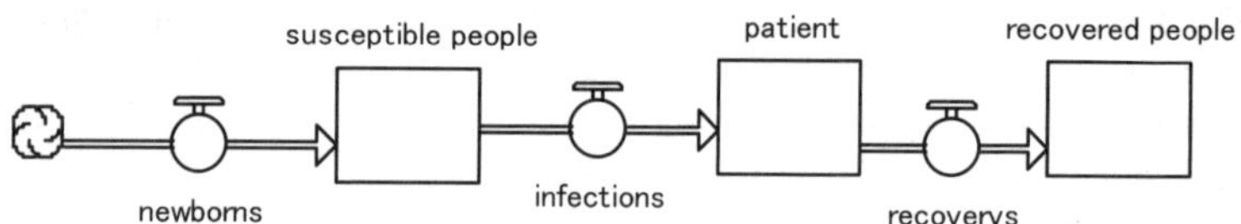

Fig. 8. All states are connected in one stock flow sequence

This kind of connection of stock flow combinations can make the diagrams themselves and equation definition processes simple and easy to understand.

2.4 Improved Translation Process

Three additional processes are shown above. Here, the whole translation process from natural language to System Dynamics models is listed. All sentences are middle language expressions such as those suggested by Takahashi [7]. Sentences express relationships between variables, X and Y, whose type (stock, flow, or auxiliary variable) are uncertain, and are conditions which should be satisfied in order to choose a model structure. One needs to try to apply patterns in the order shown below. Slashes signify alternatives

1. "X uses the same unit per time as Y. X will be/was part of Y. More/less X, more/less rapid growth/decrease of Y." This description shows that Y is a stock variable and that X is a flow variable connected to Y. See Fig. 9.

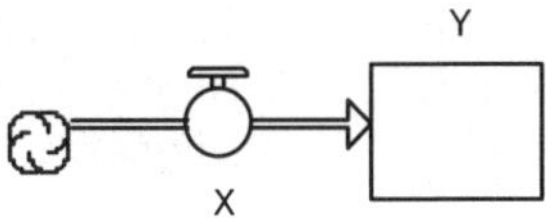

Fig. 9. "X uses the same unit per time as Y. X will be/was part of Y. More/less X, more/less rapid growth/decrease of Y."

2. "More/less X, more rapid/slower growth/decrease of Y." Y is a stock variable and X is not directly connected to Y. In addition, the variable type of X is still uncertain, but it is clear that there is another flow variable which connects X and Y. See Fig. 10.

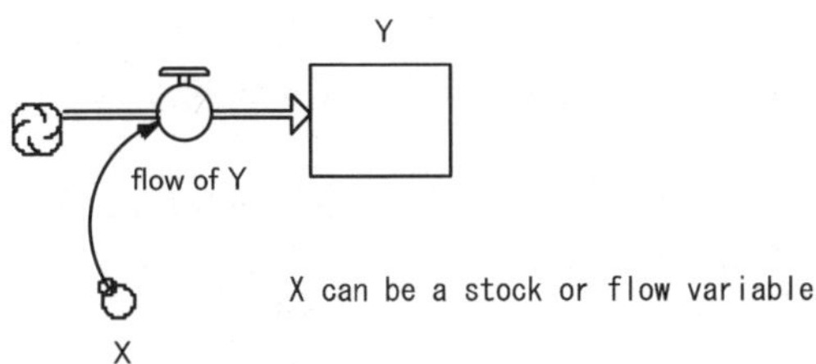

Fig. 10. "More/less X, more rapid/slower growth/decrease of Y"

3. "More rapid/slower growth/decrease of X, more rapid/slower growth/decrease of Y. X and Y use the same unit. One stays in the states expressed as X, and it can be in Y in the future." Both X and Y are stock variables and their flows are connected (from the flow of X to the flow of Y). Both of the stocks are on the same connected pipeline. See Fig. 11.

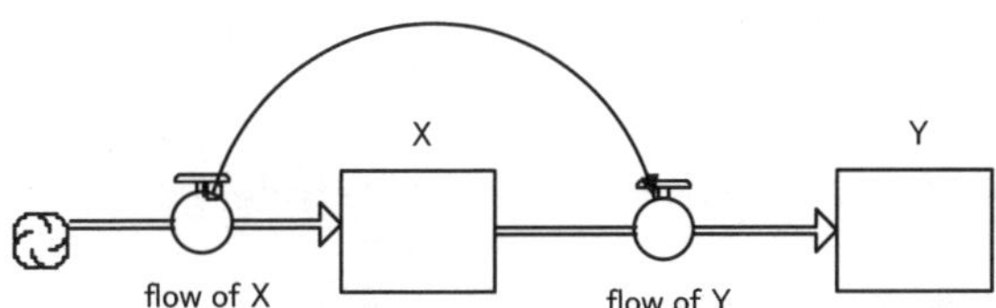

Fig. 11. "More rapid/slower growth/decrease of X, more rapid/slower growth/decrease of Y. X and Y use the same unit. One stays in the states expressed as X, and it can be in Y in the future."

4. "One is in the state expressed as X, and after that, it moves to Y." Both X and Y are stock variables and they are connected by one flow, from X to Y. Both of the stocks are on the same connected pipeline. The flow variable's definition is uncertain. See Fig.12.

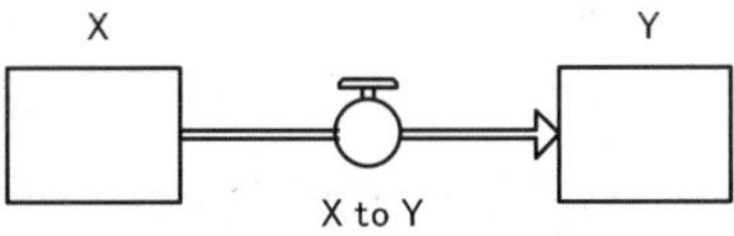

Fig. 12. "One is in the state expressed as X, and after that, it moves to Y"

5. "After X changes/increases/decreases, Y changes/increases/decreases." X is a stock variable. A causal link starts from X to Y (if Y is a flow or auxiliary variable) or to the flow of Y (if Y is a stock variable). See Fig.13.

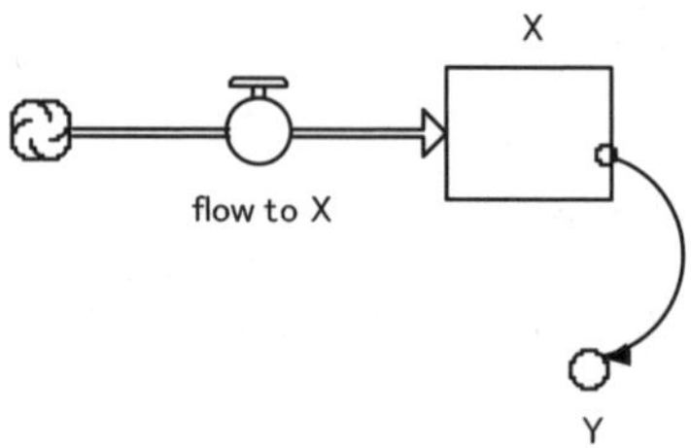

Fig. 13. "After X changes/increases/decreases, Y changes/increases/decreases"

6. "More rapid/slower growth/decrease of X, more rapid/slower growth/decrease of Y." Both X and Y are stock variables and their flows are connected (from the flow of X to the flow of Y). However, they are not on the same connected pipeline. See Fig.14.

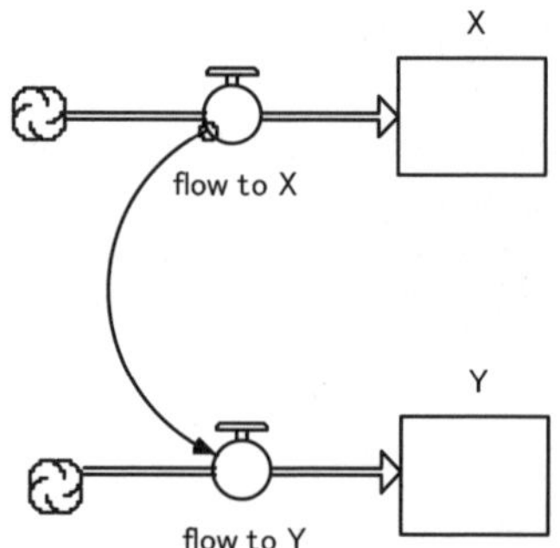

Fig. 14. "More rapid/slower growth/decrease of X, more rapid/slower growth/decrease of Y"

7. "X and Y changes/increases/decreases simultaneously." This is the same as above (6, Fig. 14) or a simple connection between flow or auxiliary variables (Fig. 15).

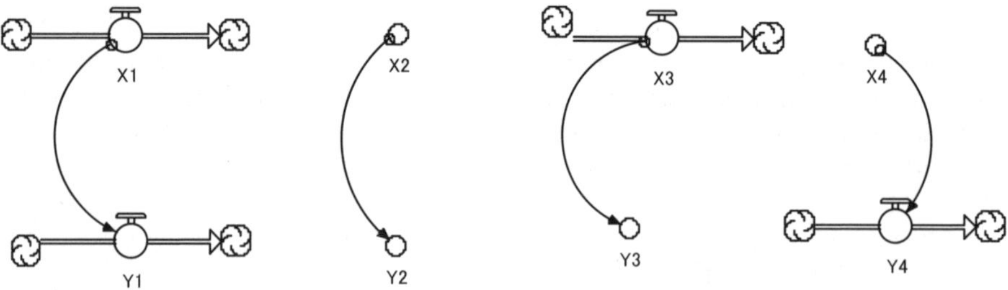

Fig. 15. "X and Y changes/increases/decreases simultaneously"

When there are still uncertain connections after using the checks above, original translation correspondences suggested by Takahashi [7] can be used.

3 Application

Anderson and Johnson [9] introduced several case studies. In one of them, they relate the story that a computer manufacturer, ComputeFast, failed to support their customers when they succeeded in acquisition of market share (Section 5). First, possible variables are chosen from the story: customers, angry customers, profit, support backlog, and support personnel. Of course, modellers can add other variables when they find later. Second, in addition to these variables, change rates need to be considered; the variables above might stock variables with flow variables. According to the story, there are additional variables: obtaining new customers, unsatisfied (customers), (support) request, supported (requests), sales, support cost, and hiring personnel. Then, the story of ComputeFast can be paraphrased using the middle language, limited syntax and vocabulary shown in the previous section, as below. The numbers in parentheses correspond to the sentence examples in Section 2.3.

[a] The customers of ComputeFast and their profit increase simultaneously. (7)

[b] More profit, more rapid growth of customers. (2)

[c] One customer is in the state expressed as customer, and after that, it moves to "angry customer." (4)

[d] More "angry customers," slower growth of customers. (2)

[e] More customers, more rapid growth of "support backlog." (2)

[f] More "support backlog," more rapid growth of "angry customers." (2)

[g] More "support personnel," more rapid decrease in profit. (2)

[h] After "support backlog" increases, "support personnel" increases. (5)

[i] After "support personnel" increases, "support backlog" decreases. (5)

There are other choices of middle language expressions; there can be other choices of combination of variables. For example, the model part [b] can be expressed as "more rapid growth of customers, more rapid grows of profit" using sentence pattern 6. All combinations should be expressed as this middle language. A variable can have possibility of having two or three choices of variable types. However, when all combinations with such a variable correspond to middle language expressions, its variable type would be determined. The important point is that using this limited syntax and vocabulary paraphrasing helps model builders to focus on correct distinctions concerning variable types and link structures.

Fig. 16 is the stock flow diagram produced from the middle language sentences, [a] to [i] above. The letters in brackets correspond to the middle language sentences.

This use of middle language is not always necessary; some can immediately determine variable types without the middle language. However, there are many cases in which one can choose two or three type of variables. Moreover, descriptive information often shows "decorative" information which is not essential for modelling. Paraphrasing to the middle language, which is also natural language, can help to clarify the problem structures by eliminating excessive information.

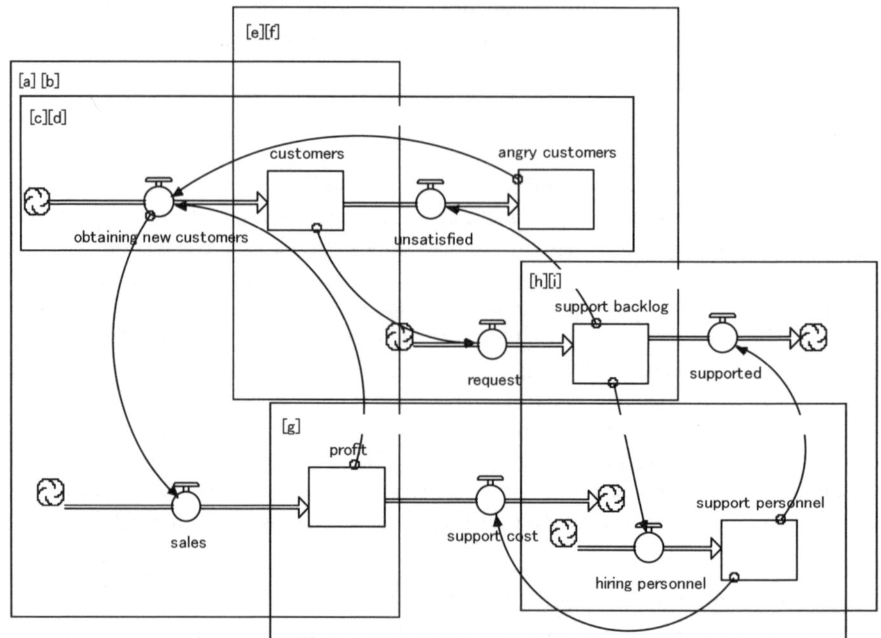

Fig. 16. After paraphrasing, model builders can connect variables automatically

4 Conclusion

This paper shows additional required information for the middle language suggested by Takahashi [7]. These additions make the middle language more complex to write. However, additions shown here can reduce the possibility of building inappropriate causal model structures. When obtaining an appropriate structure, it is easy to set parameters or functions in each variable definition using statistical or econometrical techniques. Moreover, the complexity is not in syntax but in expression. This means that syntax checks and production of middle language description for given descriptive information can be automated. Thus, the improved translation process shown here is more practical than the original process.

References

1. Jacobson, I., Christerson, M., Jonsson, P., Oevergaad, G.: Object-oriented Software Engineering. Addison-Wesley, Reading (1992)
2. Wolstenholme, E.: System Enquiry. Wiley, Chichester (1990)
3. Sterman, J.: Business Dynamics. McGraw-Hill, New York (2000)
4. Senge, P.M.: The Fifth Discipline. Doubleday (1992)
5. Sweeney, L.B., Sterman, J.: Bathtub Dynamics: Initial Results of a Systems Thinking Inventory. System Dynamics Review 16(4), 249–286 (2000)

6. Richmond, B.: An Introduction to Systems Thinking. iseesystems (1992)
7. Takahashi, Y.: Translation from Natural Language to Stock Flow Diagrams. In: Proceedings of 23rd International Conference of System Dynamics Society (2005)
8. Richardson, G.P., Pugh III, A.L.: Introduction to System Dynamics Modeling. Pegasus Communications (1989)
9. Anderson, V., Johnson, L.: Systems Thinking Basics. Pegasus Communications (1997)

Developing a Simulation Model Using a SPEM-Based Process Model and Analytical Models

Seunghun Park, Hyeonjeong Kim, Dongwon Kang, and Doo-Hwan Bae

Division of Computer Science,
Department of EECS, KAIST, Korea
{seunghun,hjkim,dwkang,bae}@se.kaist.ac.kr

Abstract. It is hard to adopt a simulation technology because of the difficulty in developing a simulation model. In order to resolve the difficulty, we consider the following issues: reducing the cost to develop a simulation model, reducing the simulation model complexity, and resolving the lack of historical data. We propose an approach to deriving a simulation model from a descriptive process model and widely adopted analytical models. We provide a method to develop simulation models and a tool environment to support the method. We applied our approach in developing the simulation model for a government project. Our approach resolves the issues by the transformation algorithms, the hierarchical and modularized modeling properties of UML and (Discrete Event System Specification) DEVS, and widely adopted analytical models.

Keywords: Simulation modeling, Hybrid simulation model, Software process modeling, SPEM, Analytical models.

1 Introduction

With the increasing interest in improving the effectiveness and efficiency of a software process, software process management becomes one of the most important issues. Organizations have tried to adopt software process simulation to manage their own software processes. However, it is difficult to adopt a simulation technology. One of the reasons is that the simulation models tend to be difficult to build and maintain [1]. In order to resolve the difficulty in developing the simulation model, we consider some key aspects of simulation modeling technology:

- **High cost for developing a simulation model**
  Developing a simulation model requires the knowledge for designing the process to be performed as well as the detailed skills for simulation tools or languages. It results in communication difficulties between the stakeholders such as a process modeler, a project manager, and simulation developer, eventually causing difficulties in technology adoption. The consistency issue between the process model to be performed, the *descriptive process*

J.L.G. Dietz et al. (Eds.): CIAO! 2008 and EOMAS 2008, LNBIP 10, pp. 164–178, 2008.

model, and the simulation model is another cost associated with the simulation model development. The consistency issue forces the software engineers to spend additional effort to develop simulation models even if the organization has a detailed process model. We need the way to reduce the development cost by automating the simulation model development as much as possible, minimizing the intervention of human agents.

- **High complexity of the process being modeled**
  The higher level of detail and fidelity to the process, greater amount of effort to build the simulation model [1]. Most simulation models are developed at once in the detailed level. It requires an amount of knowledge and experience to develop the model. We need the way to gradually develop the simulation model, lowering the abstraction level of the model.
- **Lack of the historical data**
  Quantitative equations and parameters are usually obtained from the historical data. Most of companies may suffer from collecting data that are needed for establishing the quantitative equations and parameters. Although they collect the data, people are unable to provide timely and accurate quantitative data due to the security or policy of company. We need the way to obtain the quantitative equations and parameters with less data.

A simulation model consists of the two major components: a simulation model structure and the quantitative equations and parameters. The simulation model structure provides operational guidance on the sequence of a process. The quantitative equations define functional relationships between one dependent parameter and one or more independent parameters [2]. Therefore it needs to enable a process engineer to easily and efficiently specify the simulation model structure and define the quantitative models to ease the simulation technology adoption.

We propose an approach to developing the simulation model derived from a descriptive process model and widely adopted analytical models such as COCOMO II [3]. This approach extends our previous approach by considering the definition of quantitative models [4]. We provide a method for deriving a simulation model consisting of the three steps: identifying a simulation model structure, identifying the quantitative information, and generating a simulation model. We also suggest a tool environment to support the method. We use Software Process Engineering Metamodel (SPEM) [5] and Discrete Event System Specification (DEVS)-Hybrid formalism for our research [6]. SPEM is a metamodel for defining processes and their components as a standard for process modeling [5]. SPEM can be defined as a UML Profile. It allows SPEM to gain the benefit of the expressiveness of UML. The DEVS-Hybrid simulation model is based on the DEVS-Hybrid formalism, which is an extension of DEVS formalism to the hybrid software process simulation [6].

This approach enables the simulation model development cost to be reduced by automatically transforming a descriptive process model into a simulation model, minimizing the intervention of the stakeholders. Project managers can gradually develop the hierarchical simulation model by using UML and DEVS-Hybrid simulation model. In addition, it enables the project managers to adopt the simulation technology although they do not obtain enough data to develop the quantitative equations.

The structure of this paper is as follows. In Section 2, we briefly introduce SPEM and the DEVS-Hybrid formalism as a background. Section 3 introduces the related work. Section 4 describes the method to develop a simulation model and the tool environment. Section 5 provides a case study as a validation for this approach. Section 6 summarizes the main results of this paper and gives a plan for future work.

2 Background

2.1 SPEM

The SPEM, defined by Object Management Group (OMG), is used to describe a concrete software development process or a family of related software development processes [5]. SPEM uses an object-oriented approach to modeling a family of related software processes and allows us to use UML as a notation. The SPEM is built from the SPEM Foundation package, a subset of UML 1.4, and the SPEM Extensions package, which adds the constructs and semantics required for software process modeling. The core idea of SPEM is that a software development process is the collaboration between abstract active entities, called ProcessRoles, which perform operations, called Activities or Steps, on concrete, tangible entities, called WorkProducts.

2.2 DEVS-Hybrid Formalism

DEVS-Hybrid simulation model is based on the DEVS-Hybrid formalism, which is extended to accommodate the hybrid characteristics of software development process [6]. DEVS-Hybrid simulation model uses the system dynamics modeling to convey the details concerning the activity behaviors and managerial policies, while the discrete event modeling controls the activity start/finish and sequence. DEVS-Hybrid can represent the discrete activities explicitly and consistently with the continuously varying project environments by fully incorporating the feedback mechanism of the system dynamics. Similarly DEVS [7], DEVS-Hybrid simulation model has two kinds of models to represent systems. One is an atomic model and the other is a coupled model. While the coupled model is the same as DEVS, the atomic model is extended. The DEVS-Hybrid formalism for the atomic model as follows [8]:

$$\textit{DEVS-Hybrid} = \langle X, Y, Y^{phase}, S, \delta_{ext}, \delta_{int}, C_{phase}, \lambda, ta \rangle$$

where:
X is a set of input values. Y is a set of output values. Y^{phase} is a set of output values, which is the phase event triggered by phase event condition function (C_{phase}). S is the set of states. $\delta_{ext} : Q \times X \rightarrow S$ is the external transition

function, where $Q = \{(s,e)|s \in S, 0 \leq e \leq ta(s)\}$ is the total state set, e is the time elapsed since last transition. $\delta_{int} : S \rightarrow S$ is the internal transition function. $C_{phase} : Q \times X \rightarrow Bool$ is the phase event condition function for conditioning the execution of the phase event. $\lambda : S \rightarrow Y$ is the output function. $ta : S \rightarrow R^+_{0,\infty}$ is the set positive reals between 0 and ∞.

An atomic model can stay only in one state at any time. The maximum time to stay in one state without external event is determined by $ta(s)$ function. δ_{ext} distinguishes the input into discrete or continuous one, and the input follows two different processing path after that. When an atomic model is in a state $0 \leq e \leq ta(s)$, it changes its state by δ_{ext} if it gets an external discrete event. If the model gets the external continuous input, which contains the flow(rate) and stock variables, the model updates the stock variables by using the equation as shown in [6]. If possible remaining time in one state is passed, it generates output by λ and changes the state by δ_{int}. If a phase event condition function becomes true, the phase output event(Y^{phase}) occurs.

A coupled model is constructed by coupling the atomic models or other coupled models. Through the coupling, the output events of one model are converted into input events of other models. In DEVS theory, the coupling of DEVS models defines new DEVS models (i.e., DEVS is closed under coupling) and then complex systems can be represented by DEVS in a hierarchical way. In the DEVS-Hybrid coupled model, the two atomic models are connected by input and output ports internally, which is called Internal Coupling (IC). An atomic model is connected by external input which is called External Input Coupling (EIC) and external output which is called External Output Coupling (EOC) with a coupled model.

The DEVS-Hybrid simulation model is executed by DEVSim++, which is a C++ based DEVS simulation environment [9].

3 Related Work

There are researches to develop a simulation model more systematically and efficiently. Pfahl et al. developed a methodology for integrated measurement, modelling and simulation (IMMoS) [2]. IMMoS helps industrial practitioners to build system dynamics (SD) models efficiently and effectively. IMMoS integrates system dynamics modeling with goal-oriented measurement and descriptive process modeling. In order to integrate the descriptive process model into a SD model, authors map the descriptive process model elements to SD model flow graphs representation. Quantitative models by goal-oriented measurement are used to define start conditions, implicit management decision rules, and explicit management decision rules of SD models. This approach provides the systematic methodology to develop a simulation model based on the descriptive process model. However, the approach still requires simulation model developers to share the knowledge about each model which causes the communication difficulty. It also does not provide the way to ensure the consistency between the descriptive process model and the simulation model.

Raffo et al. proposed the generalized process simulation model (GPSM) concepts [17]. Authors introduce generic building blocks which enable modularization, and reuse of components. A GPSM is developed by tailoring the existing generic GPSM blocks. This work makes it easy to reduce the time and cost of building a simulation model by reusing the modularized building blocks. However, it still requires the communication overhead between stakeholders to select and tailor the building blocks. In addition, there is no guidance to develop the quantitative equations.

Donzelli proposed the hybrid simulation modeling approach [11]. He describes the simulation model structure by using a discrete-event queuing network. He uses the analytical and the continuous modeling methods to describe the dynamic behavior of each activity. This approach allows the simulation model developers to utilize the predefined models. However, we should describe the process model in queuing network. It makes the process model to be difficult to develop and understand. As the result, the communication cost is increased.

4 A Method for Developing a Simulation Model and a Tool Environment

Figure 1 illustrates the overview of our approach to derive a DEVS-Hybrid simulation model from a SPEM-based process model and the analytical models.

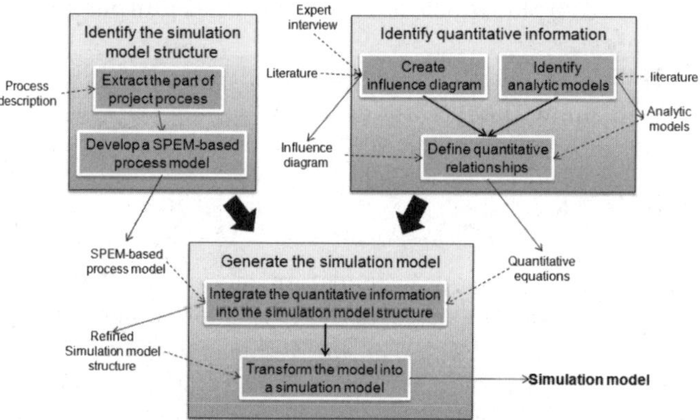

Fig. 1. The procedure of the simulation model development

At first, we select the part of the process to be simulated and develop a process model using UML representation based on SPEM. The SPEM-based process model represents the simulation model structure. And then we define quantitative relationships by using an influence diagram and the analytical models widely adopted in the software community. The quantitative relationships are integrated into the simulation model structure. Finally, the integrated simulation model structure is automatically transformed into a DEVS-Hybrid simulation model.

4.1 Identifying the Simulation Model Structure

Simulation models can be developed at different levels of scope and depth to suit organization's needs [1]. We should decide which part of a process is simulated depending on the scope and purpose of simulation. For example, when we want to analyze the impact of requirements change on the schedule or effort, we may consider the directly and indirectly affected parts of a process. We may also consider the whole process to identify the impact of the change on the process performance. After deciding the part of a process to be simulated, we model the part of a process using UML representation.

The UML profile for SPEM gives benefits of using UML diagrams to present different perspectives of a software process model [5]. We use the two UML diagrams among them: Use case diagram and Activity diagram. The examples are presented in Section 4. Use case diagram describes the assignment of ProcessRoles to Phases or Activities and represents the Work Breakdown Structure (WBS) of the process. For example, a Phase can include many Activities. Stereotyped ≪include≫ relation is used to specify the relationship between a Phase and the Activities which comprise the phase. Using the use case diagram, we can hierarchically construct the coupled model of a DEVS-hybrid simulation model.

Activity diagram represents the sequence of activities with their input and output WorkProducts. Activity diagram can also illustrate the behavior of an activity with the ActionStates which represent Steps of an Activity. In general, the statechart diagram illustrates the behavior of an activity. SPEM specification defines that Step should be used only in the activity diagram. In this usage of the activity diagram, it defines how to change the state based on the input message and generates outputs after entering a specific state like the statechart diagram in order to calculate the quantitative parameters. All the parameters defined in the activity diagram are used as attributes of the atomic model and the attributes are instantiated by the scenario of DEVS-hybrid simulation model.

4.2 Identifying Quantitative Information

We should identify the causal relationships to represent the dynamic behavior of a simulation model. A cause-effect or influence diagram is widely used to describe the relationships. These diagrams describe the factors which typically change the behavior of other project factors and the causal relationships between the factors. The causal relationships can be obtained by analyzing the literature or expert's opinion.

Several approaches tried to apply well-known analytical models to a simulation technology. We have analyzed the simulation models which use the analytical models. Abdel-Hamid and Madnick use COCOMO to obtain the initial estimates for the completion time, the number of tasks to be accomplished, and the required effort and manpower [10]. The initial estimates provide the baseline performance of the process. Donzelli et al. apply COCOMO equations to estimate the size of the output artifacts and the effort and delivery time required to perform the corresponding tasks [11]. They also apply the Rayleigh model to predict the staffing profile. Madachy develops a system dynamics model of an inspection-based process to evaluate the effects of inspections [12]. The development rate of each phase is constrained by the manpower allocation and current productivity which are derived from COCOMO. The modified Rayleigh staffing curve dynamically distributes the effort based on the development progress compared to the development schedule. These approaches give us insights into the possibility of using analytical models in a simulation technology.

We have provided the way to use COCOMO II to estimate the effort, schedule, and quality of the software development project when the project environment is dynamically changing [8]. We decomposed the time-aggregated effort and schedule estimates of COCOMO II into the time-continuously varying development rate of each phase by using the effort and schedule distribution data. The effort and schedule estimates provided by COCOMO II are time-aggregated in that they just show the end state of the project. To dynamically simulate the development process, the model needs a small-time incremented development rate, which also needs to be dynamically changed during the course of a project according to the effects of the dynamic environment.

4.3 Generating the Simulation Model

The quantitative equations and parameters are incorporated into the activity diagrams which describe the behavior of each activity. The keyword *do* represents an ongoing activity that is performed as long as the modeled element is in the state. Using the keyword *do*, we specify the equations and parameters. For example, if we want to calculate the development rate of a specific activity, we describe the equation for the development rate with the expression, "*do* DevelopmentRate_A = ManpowerRate_A*Productivity_A".

After integrating the quantitative equations and parameters into the SPEM-based simulation model structure, we can transform the simulation model structure into a DEVS-hybrid simulation model. We propose the algorithms to automate the transformation: the structural transformation algorithm and the behavioral transformation algorithm.

The structural transformation algorithm uses a use case diagram and the activity diagram describing the overall sequence of all the activities in a process. Figure 2 shows the structural transformation algorithm. We assume the activities in the activity diagram are one to one mapped to the use cases stereotyped with ≪Activity≫ in the use case diagram. The use cases stereotyped with ≪Phase≫ are transformed into the coupled models. Activities included in a phase are transformed into the corresponding atomic models. The interactions between the activities in the same phase are transformed into ICs of the phase. Inputs from the activities in the different phases are transformed into the input variables of the coupled model and EICs between the atomic models which receive the inputs and the coupled model including the atomic models. Similarly, outputs toward the activities in the different phases are transformed into EOCs and output variables of the coupled model. One of the output variables is transformed into the phase output variable.

```
Input: a UseCase diagram, an Activity diagram describing the overall structure
Output: Coupled model

- UCi, UCm, UCn : use cases in the UseCase diagram
- ACi, ACm, ACn : activities in the Activity diagram
- S : a set of use cases which are the targets of include relation associated with UCi
- T : a set of activities receiving the work product which ACj sends
-Wp : the work products between ACj and ACm
-Wq : the work products between ACn and ACj
- Z : a set of activities sending the work product to ACj

For all use cases UCi in the UseCase diagram

if UCi's stereotype = "Phase" or "WorkDefinition"
{
    UCi = Coupled Model

    For all UCj ∈ S {
        ACj = Ucj

        /* Construct outgoing relations from an activity */
        For all ACm ∈ T {
            UCm = ACm
            If (UCm ∈ S)        // if two activities are performed in the same phase
                IC = Wp
            Else {              // if two activities are performed in the different phase
                EOC = Wp
                If (!(Wp ∈ Y or Yphase))   // Wp does not already exist as an output
                    Y = Wp or Yphase = Wp
            }
        }

        /* Construct incoming relations to an activity */
        For all ACn ∈ Z {
            UCn = ACn
            If (ACn ∈ S)               // if two activities are performed in the same phase
                IC = Wq
            Else {              // if two activities are performed in the different phase
                EIC = Wq
                If (!(Wq ∈ X))   // Wq does not already exist as an input
                    X = Wq
            }
        }
    }
}
Else if UCi's stereotype = "Activity"
    UCi = Atomic Model
```

Fig. 2. The structural transformation algorithm

Figure 3 shows the behavioral transformation algorithm. All ActionStates representing the Steps are transformed into the states of the atomic model and all external events are transformed into input variables. Guard condition on a transition is used to check whether an activity is complete or not and transformed into the phase event condition function. The action with a guard condition is transformed into the phase output variable. The transition triggered by an external event is transformed into the external transition function. On the other hand, the transition triggered when the specified time is elapsed is transformed into the internal transition function and the action on the transition is transformed into the output function.

We need to confirm that the transformation algorithms operate correctly. We validate the transformation algorithms with the criteria proposed in [13]: termination and behavioral equivalence. The termination means that the transformation algorithm should be applied finitely. The number of applications of the structural transformation algorithm and the behavioral transformation algorithm are limited to the number of use cases. Therefore, the transformation algorithms in our approach are applied finitely. The behavioral equivalence means the behavior of a source model is preserved in the transformed model. The behavior of models is described in the activity diagrams and atomic models. Table 1 shows

```
Input: a UseCase diagram, Activity diagrams describing the behavior of the activities
Output: Atomic model
- UC_i : a use case in the UseCase diagram
- S_devs : states in an atomic model
- S_step : ActionStates in the Activity diagram
- E : external events on all the transitions in an Activity diagram
- A : actions on all the transitions in an Activity diagram
- AE : an action associated with the TimeEvent

For all use cases in the UseCase diagram
if UC_i's stereotype = "Activity"  {
    For the activity diagram which is mapped to UC_i  {
        S_devs = S_step
        X = E
        if (guard condition on a transition)  {
            Y_phase = A
            C_phase = guard condition
        }
        Else
            Y = A
        For all events e in the Activity diagram  {
          if (e == "after")  {  // if the event is a TimeEvent
             δ_int = (the source state, the target state)
             λ = (the source state, AE)
             ta = time of "after"
          }
          Else  {
             δ_ext = (the source state, the target state, e)
             ta = infinity
          }
        }
    }
}
```

Fig. 3. The behavioral transformation algorithm

Table 1. The mapping between the elements of activity diagram and atomic model

The elements of an activity diagram	DEVS-Hybrid
Event: E	Input value: X
Action: A	Output value : Y, Phase output value : Y^{phase} Output function (λ)
ActionState: S	State: S
Transition: T	External transition function (δ_{ext}) Internal transition function (δ_{int})
TimeEvent: *after*	Time advance function(ta)
Guard condition	Phase condition function (C_{phase})

the mapping between the elements of an activity diagram and an atomic model. The mapping is deduced from the fact that the two models are based on the state machine. As shown in Table 1, the behavioral properties of every element of an activity diagram is one to one mapped to atomic model of DEVS-Hybrid simulation model. The behavior of the source model, therefore, is preserved in the transformed model.

4.4 Tool Environment to Support the Method

The proposed method can be automatically performed by supporting tools. Figure 4 shows the tool environment to support the method we proposed. This

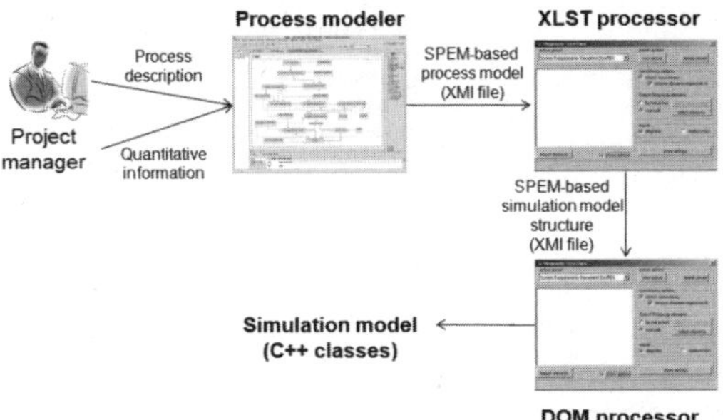

Fig. 4. Tool environment

tool environment can cover the overall procedure we proposed. The process modeler supports the process modeling using UML representation based on SPEM. We can develop the simulation model structure and integrate the quantitative information using the process modeler. The process modeler exports a XML Metadata Interchange (XMI) file for an interchanging model in a serialized form. Extensible Stylesheet Language Transformations (XSLT) processor automatically transforms the exported XMI file into a XML file by applying the transformation algorithms. Document Object Model (DOM) processor finally generate DEVSim classes executed with DEVSim++ engine.

5 A Case Study

In this section, we provide a case study for validating our approach in the industry. We develop the simulation model supporting for deciding the acceptable requirements creeping. The simulation model can help project managers to analyze how much requirements can be added or changed in given constraints on schedule or cost. We describe the procedure for developing the simulation model and show the result of simulation. The project introduced in this paper is the same as in [4]. The project type is the new development of information system. The initial size of the project is 30KLOC. The lifecycle model is the typical waterfall model. There has been the requirements creeping as the following pattern: Design (77%), Code (13%), Test (10%).

5.1 The Descriptive Process Model Represented by UML

There was several subprocesses in the project. We selected the process related to the business process management system. The use case diagram shown in Figure 5 describes the part of the decomposition of the process into several phases and activities applied to the project. The process component "Develop Business Process Management (BPM)" consists of the four phases: "Requirements", "Design", "Code", and "Test". Each phase also consists of several activities such as "Identify the customer requirements". Figure 6 shows the inputs, outputs, and sequences of activities. The swimlane represents ProcessRoles assigned to each

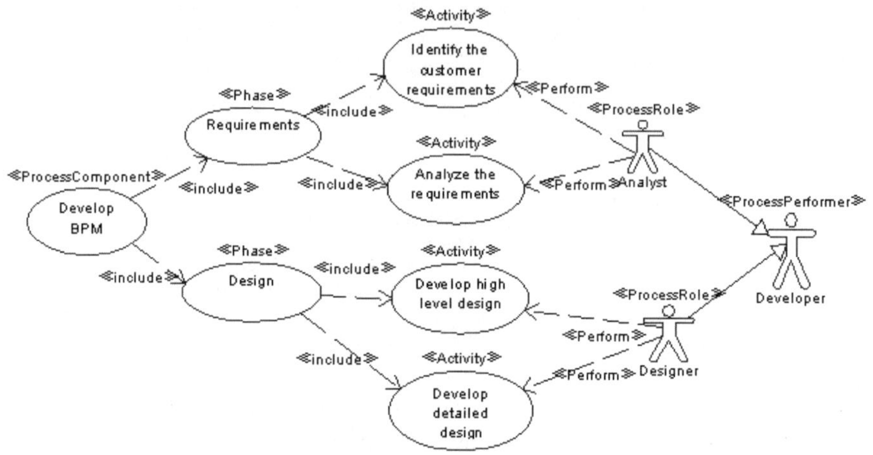

Fig. 5. The part of the use case diagram of the descriptive process model

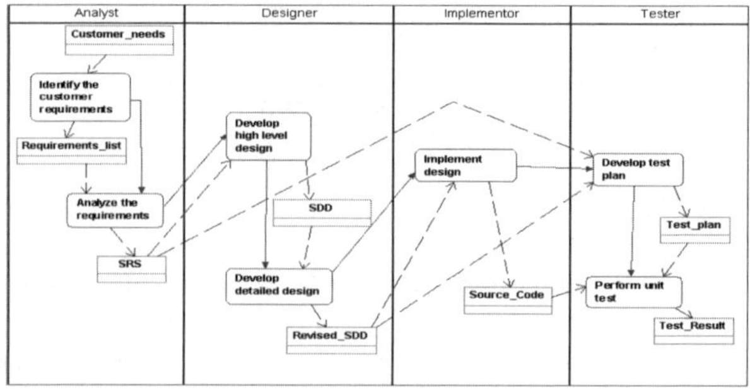

Fig. 6. The part of an activity diagram of the descriptive process model

activity. This activity diagram gives the overall sequence of the process. Figure 7 describes the behavior of the activity "Identify the customer requirements". This activity diagram focuses on the timing of process steps. The activity consists of the two steps: "Interview with customer", and "Develop requirements list". When the activity "Identify the customer requirements" receives the work product "Customer_needs" as an input, the activity enters to the state "Interview with customer" and calculates the effort and the amount of the performed work. After entering the state "Develop requirements list", TimeEvent occurs after the amount of time given to the parameter "TimeForDevelopList" elapses. With this event, if customer needs are identified over than 90%, then the work product "Requirements_list" is produced as an output and transferred to the next activity "Analyze the requirements".

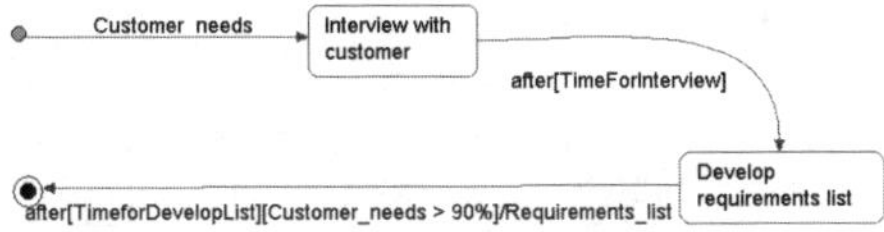

Fig. 7. The activity diagram for the activity "Identify the customer requirements"

5.2 The Quantitative Information

The quantitative equations and parameters are defined based on the COCOMO II proposed in [8]. For example, we can calculate the development rate of each activity by the equation, $DevelopmentRate = ManpowerRate \times Productivity$. The productivity of each activity is the amount of the job size that can be performed by the unit effort in the activities. The development rate of each phase is the product of the manpower rate and productivity of each activity, which shows how much work is processed in the unit time. The ManpowerRate of each activity is the rate of how many personnel is put into the activity in the unit time. This can be obtained by multiplying the two values: average full-time equivalent software personnel to complete the project, which is derived by dividing the total effort by the total schedule and the rate of effort(effort(%)/schedule(%)) required in the activity.

We adopt the use case point method to calculate the available number of requirements [14], [15]. We assume one use case represents one requirement. Given schedule or cost constraints, we can know the size of available requirements creeping in lines of code unit. We use the size as an input of COCOMO II. Equation (1) shows how to calculate the number of use cases.

$$\sum UC_i \times P_i = A \times (Size)^B \times \prod EM. \tag{1}$$

UC_i represents the number of use cases which can be developed additionally. P_i means the required effort per a use case. P_i has different values based on the size of software, but we assume it has same value in one project.

5.3 DEVS-Hybrid Simulation Model

Figure 8 shows the part of the derived simulation model by applying the transformation algorithms to the SPEM-based process model including the quantitative information.

Each phase is transformed into the corresponding coupled models and includes the activities which are transformed into the corresponding atomic models. For example, the phase "Requirements" includes the two activities in the use case diagram: "Identify customer requirements" and "Analyze the requirements". The coupled model "Requirements" also includes the two atomic models in the DEVS-Hybrid simulation model: "Iden_Cust_Reqs", and "Analy_Reqs". The name of those atomic models has the compressed form of the activity "Identify customer requirements" and the activity "Analyze the requirements", respectively, because of the convenience of displaying. The work products of the

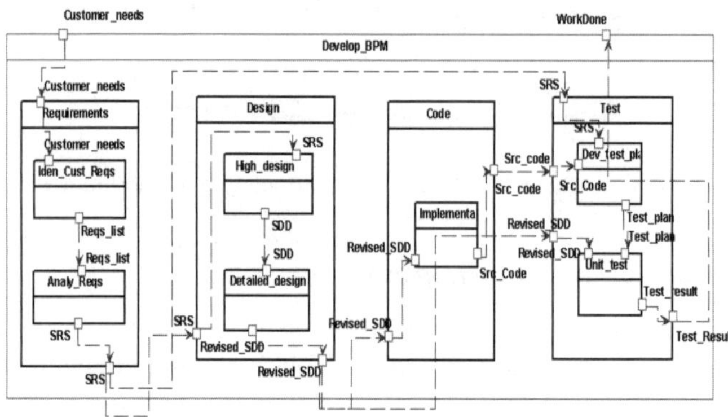

Fig. 8. The part of the transformed DEVS-Hybrid simulation model

activities are transformed into the corresponding EICs, EOCs, and ICs. The work product "Requirements_list", for example, is sent from the activity "Identify customer needs" and received by the activity "Analyze requirements" in the descriptive process model. Because the two activities are in the same phase, the work product should be transformed to IC followed by the structural transformation rule. The IC between the atomic model "Identify Customer requirements" and the atomic model "Analyze requirements" is "Requirements_list". The transformation, therefore, is correctly performed.

The simulator which is implemented using the DEVSim++ and available at [16] for our research. We can design various simulation scenarios by changing the upper limit of the schedule or cost. The simulation tool can show the simulation results with the numerical values and a chart. Figure 9 shows the simulation results under various cost upper limits. When we set the cost upper limit as 10% of the initial cost, 24 use cases or requirements can be acceptable at the maximum. After the cost upper limit reaches 30% of the initial cost, few requirements can be accepted. The project manager decided that he could accept less than 40 requirements because accepting over than 40 requirements is not cost-effective.

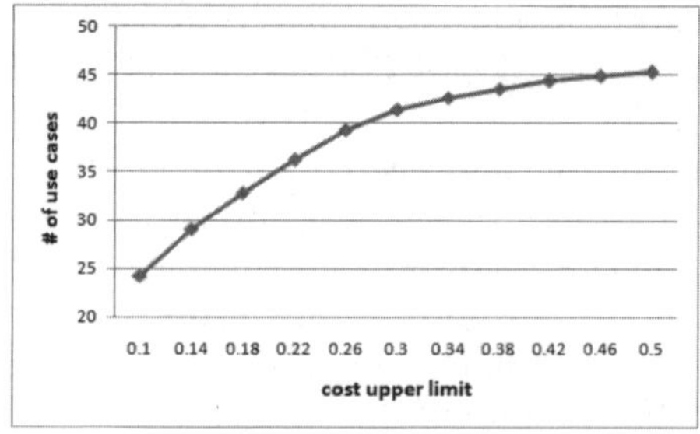

Fig. 9. The number of acceptable use cases

6 Conclusion and Future Work

We proposed an approach to developing simulation models by deriving a simulation model from the predefined models. This approach can resolve the issues that we raised.

- **High cost for developing a simulation model**
  Once a descriptive process model is developed, a simulation model can be automatically generated from the descriptive process model. It minimizes the interventions of stakeholders resulting in reduction of communication cost. It also ensures the consistency because the descriptive process model is directly transformed into a simulation model.
- **High complexity of the process being modeled**
  UML supports the hierarchical modeling with several diagrams. DEVS-Hybrid formalism also supports the hierarchical modeling with the coupled models, and the atomic models. We can develop the overall structure of a simulation model at first and then build up the detailed behavior.
- **Lack of the historical data**
  There are many predefined analytical models such as COCOMO. Those analytical models are developed by the historical data from a variety of organizations. By adopting the analytical models, we can supplement the lack of historical data. But we need to calibrate the models before adopting.

As a future work, we will apply our approach to a variety of methods for quantitatively analyzing a descriptive process model. At first we will apply our approach to analyzing the impact of process changes during process enactment in Process-centered Software Engineering Environments (PSEEs). A PSEE contains the descriptive process model to be enacted. By transforming the process model to be changed into a simulation model and simulating the model, it can help a project manager decide whether the change is adopted or not. We will apply our approach to process construction or process tailoring by using the existing process modules used in constructing a simulation model.

Acknowledgments. This research was supported by the MIC(Ministry of Information and Communication), Korea, under the ITRC(Information Technology Research Center) support program supervised by the IITA(Institute of Information Technology Advancement)(IITA-2008-(C1090-0801-0032)). This work was also partially supported by Defense Acquisition Program Administration and Agency for Defense Development under the contract.

References

1. Raffo, D., Spehar, G., Nayak, U.: Generalized Simulation Models: What, Why and How? In: ProSim 2003, Oregon (2003)
2. Pfahl, D., Ruhe, G.: IMMoS: A Methodology for Integrated Measurement, Modelling and Simulation. Software Process Improvement and Practice 7(3-4), 189–210 (2002)

3. Boehm, B.: Software Cost Estimation with COCOMO-II. Prentice-Hall, Englewoord Cliffs (2006)
4. Park, S., Choi, K., Yoon, K., Bae, D.: Deriving Software Process Simulation Model from SPEM-based Software Process Model. In: APSEC 2007, Nagoya, Japan, pp. 382–389 (2007)
5. Software Process Engineering Metamodel Specification, Version 1.1, OMG Document formal/05-01-06 (2005)
6. Choi, K., Bae, D., Kim, T.: An approach to a hybrid software process simulation using DEVS formalism. Software Process Improvement and Practice 11(4), 373–383 (2006)
7. Zeigler, B., Pracehofer, H., Kim, T.: Theory of Modeling and Simulation, 2nd edn. Academic Press, New York (2000)
8. Choi, K.: Hybrid Software Process Simulation Modeling for Analyzing Software-Intensive System Acquisition. Ph.D Dissertation, KAIST (2007)
9. Kim, T.: DEVSimHLA v2.2.0 Developer's Manual, KAIST (2004)
10. Abdel-Hamid, T., Madnick, S.: Software Project Dynamics: An Integrated Approach. Prentice-Hall, Englewood Cliffs (1991)
11. Donzelli, P.: A Decision Support System for Software Project Management. IEEE Software 23(4), 67–75 (2006)
12. Madachy, R.: System dynamics modeling of an inspection-based process. In: Proceeding of the 18th ICSE, pp. 376–386 (1996)
13. Lara, J., Taentzer, G.: Automated Model Transformation and its Validation Using AToM3 and AGG. In: Blackwell, A.F., Marriott, K., Shimojima, A. (eds.) Diagrams 2004. LNCS (LNAI), vol. 2980, pp. 182–198. Springer, Heidelberg (2004)
14. Smith, J.: The estimation of effort based on use cases. Rational Software, white paper (1999)
15. Mohagheghi, P., Anda, B., Conradi, R.: Effort Estimation of Use Cases for Incremental Large-Scale Software Development. In: ICSE 2005, St Louis, USA, pp. 303–311 (2005)
16. Project Management Simulator, `http://spic.kaist.ac.kr/~selab/html/Simulator.zip`
17. Raffo, D., Nayak, U., Wakeland, W.: Implementing Generalized Process Simulation Models. In: ProSim 2005, St. Louis (2005)

Formal Modeling and Discrete-Time Analysis of BPEL Web Services

Radu Mateescu[1] and Sylvain Rampacek[2]

[1] INRIA / VASY, Centre de Recherche Grenoble – Rhône-Alpes, France
Radu.Mateescu@inria.fr
[2] LE2I, Université de Bourgogne, Dijon, France
Sylvain.Rampacek@u-bourgogne.fr

Abstract. Web services are increasingly used for building enterprise information systems according to the Service Oriented Architecture (SOA) paradigm. We propose in this paper a tool-equipped methodology allowing the formal modeling and analysis of Web services described in the BPEL language. The discrete-time transition systems modeling the behavior of BPEL descriptions are obtained by an exhaustive simulation based on a formalization of BPEL semantics using the Algebra of Timed Processes (ATP). These models are then analyzed by model checking value-based temporal logic properties using the CADP toolbox. The approach is illustrated with the design of a Web service for GPS navigation.

Keywords: Web services, formal specification, model checking, exhaustive simulation, process algebra.

1 Introduction

Information systems present in companies and organizations are complex software artifacts involving concurrency, communication, and coordination among various applications that exchange data and participate to business processes. *Service Oriented Architecture* (SOA) [15] is a state-of-the-art methodology for developing information systems by structuring them in terms of services, which can be distributed and composed over a network infrastructure to form complex business processes. Web services are a useful basis for implementing business processes, either by wrapping existing software or by creating new functionalities as combinations of simpler ones. BPEL (*Business Process Execution Language*) [14] is a standardized language of wide industrial usage for describing abstract business processes and detailed Web services. It allows to capture both the behavioral aspects (concurrency and communication) and the timing aspects (duration of activities) of Web services.

The BPEL language allows to create Web services either from scratch, or as the composition of existing sub-services, which can be invoked sequentially (one at a time) or concurrently (several ones at the same time). Each Web service described in BPEL can be used as a sub-service by other Web services (described

J.L.G. Dietz et al. (Eds.): CIAO! 2008 and EOMAS 2008, LNBIP 10, pp. 179–193, 2008.

in BPEL or not), thus enabling a hierarchical construction of complex Web services. A BPEL business process is defined by a workflow consisting of various steps, which correspond internally to algorithmic computations (possibly with time constraints) and externally to message-passing interactions with a client. Business processes are typically built upon existing Web services (although this is not mandatory), each one being specialized for carrying out a particular task. These sub-services are invoked every time a specific information is needed during a step of the workflow; therefore, a business process is not simply the set of sub-services it is built upon, but acts as an orchestrator of these sub-services in order to provide newly added functionalities.

The conjunction of concurrency and timing constraints makes business processes complex and requires a careful design in order to avoid information losses and to obtain a satisfactory quality of service. In this context, formal modeling and analysis techniques available from the domain of concurrent systems allow to improve the quality of the design process and to reduce the development costs by detecting errors as soon as possible during the business process life cycle. These techniques can operate successfully on languages equipped with a formal semantics definition, from which suitable models can be constructed and analyzed automatically.

In this paper, we propose a tool-supported approach for the formal modeling and analysis of business processes and Web services described in BPEL. Our approach consists of the following ingredients: the definition of a formal semantics of BPEL in terms of process algebraic rules, taking into account the discrete-timing aspects [11,12]; the automated generation of models (state/transition graphs) from the BPEL specifications using an exhaustive simulation based on the formal semantics rules, implemented in the WSMOD tool; and the analysis of the resulting models by using standard verification tools for concurrent systems, such as CADP [9]. We illustrate the application of this approach to the design and discrete-time analysis of a Web service for GPS navigation.

Related work. The modeling and analysis of Web services benefits from a considerable attention in the research community. The WSAT tool proposed in [5,6] gives to Web service designers the possibility of verifying LTL properties on BPEL business processes by applying the SPIN model checker. Each BPEL process is transformed into a PROMELA model (via a pattern) and connected to other processes in the description. This work covers only the untimed aspects of BPEL.

Another approach, proposed in [28], uses the CRESS (*Chisel Representation Employing Systematic Specification*) notation for specifying the untimed behavior of Web services. CRESS descriptions are translated into the formal description technique LOTOS [13] and analyzed with dedicated tools, such as TOPO, LOLA or CADP. A direct translation from BPEL to LOTOS is given in [4], enabling the use of the aforementioned tools for analyzing the untimed behavior of Web services. BPEL was also used as target language for producing executable Web services from LOTOS specifications [2,25]; this allows to combine the advantages of the formal verification using CADP and of the deployment and execution features of BPEL.

Compared to existing work, our approach differs in the following respects: it is based on a translation of BPEL directly into state/transition graphs, without using an intermediate language such as PROMELA or LOTOS, thus being potentially more efficient; and it handles not only the behavioral, but also the discrete-time aspects of BPEL descriptions.

Paper outline. Section 2 presents our methodology and software platform for modeling and analyzing BPEL descriptions. Section 3 describes the GPS Web service case-study and its analysis using the platform. Finally, Section 4 gives some concluding remarks and directions for future work.

2 Modeling and Analysis Approach

Web services can be seen as complex distributed systems that communicate by message-passing. Therefore, their design methodology can be naturally supported by the formal modeling and analysis techniques stemming from the domain of concurrent systems. To apply these techniques, it is necessary to represent the dynamic behavior of Web services in a formal, non-ambiguous manner.

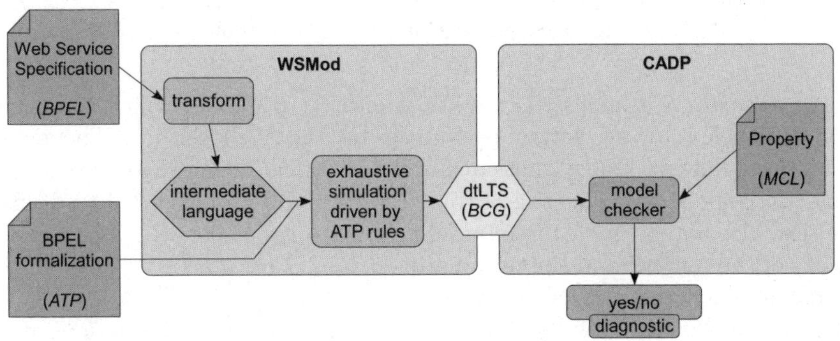

Fig. 1. Platform for Web service modeling and analysis

The approach we propose for the modeling and analysis of Web services described in BPEL is illustrated in Figure 1. Our software platform consists roughly of two parts, described in the sequel: the BPEL descriptions are first translated into discrete-time LTSs using the WSMOD tool, and are subsequently analyzed using the CADP verification toolbox.

2.1 Translation from BPEL to Discrete-Time LTSs

The behavior of a Web service comprises not only the concurrency and communication between its various constituent activities, but also the delay of response of the service. These aspects can be modeled using DTLTSs (*discrete-time*

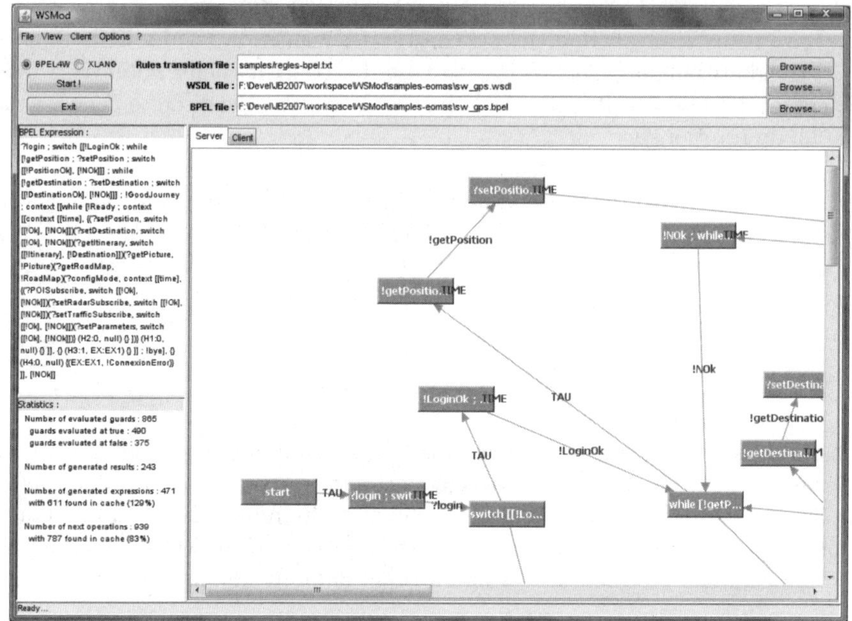

Fig. 2. Screenshot of the WSMod tool

Labeled Transition Systems), i.e., state/transition graphs in which every transition is labeled by an action performed by the Web service. The actions are of the following kinds: emissions and receptions of messages, prefixed by '!' and '?', respectively; elapsing of time, represented by the symbol χ (discrete-time tick, also noted *time*); the internal action τ (or *tau*) denoting unobservable activity of the service; and the terminating action $\surd$ (or *done*), which is the last internal action that a service can do.

The global behavior of the Web service (and therefore, the actions it can perform) is obtained by an exhaustive simulation of the Bpel description, performed by the WSMod tool (see a screenshot in Figure 2), which is able to handle both discrete [11] and continuous [12] time representations. WSMod takes two different inputs (see Figure 1):

- A Web service description in Bpel [14], a standardized language allowing to specify the behavior of business processes. Bpel supports two different types of business processes: *executable* processes specify the behavior of business processes in full detail, such that they can be executed by an orchestration engine; and *abstract* business protocols specify the public message exchanges between the service and a client (i.e., excluding the message exchanges which take place internally, e.g., during invocations of sub-services).
- A formal representation of the Bpel semantics, based on the Algebra of Timed Processes (Atp) [24], which specifies using operational rules how the model of the business process behavior is generated. Depending on the time

representation chosen, the resulting model is either a DTLTS, or a timed automaton (TA) [1]. An excerpt of the ATP rules formalizing the BPEL semantics in discrete-time is shown in Table 1. For example, the process "time" can only elapse time (represented by the χ action), and the process "receive" or "reply" can send or receive a message (first rule) or elapse time too (second rule).

To generate the model representing the behavior of the input BPEL description, WSMOD performs an exhaustive simulation guided by the operational ATP rules. The tool is also able to synthesize automatically the model of an adapted client interacting with the Web service, whose behavior complies with that of the service as regards emission and reception of messages, time elapsing, etc. In this study, we focus only on the Web service model generation feature of WSMOD.

2.2 Analysis of Discrete-Time LTSs

Once the DTLTS model of the BPEL specification under design has been obtained, it can be analyzed by using standard tool environments available for concurrent systems. For our purpose, we use the CADP (*Construction and Analysis of Distributed Processes*) toolbox [9] dedicated to the formal specification and verification of concurrent asynchronous systems. CADP accepts as input specifications written in process algebraic languages, such as LOTOS [13], FSP [18,26] or CHP [19,27], as well as networks of communicating automata given in the EXP language [16]. These formal specifications are translated by specialized compilers into labeled transition systems (LTSs), i.e., state spaces modeling exhaustively the dynamic behavior of the specified systems. LTSs are the formal model underlying the analysis functionalities offered by CADP, which aim at assisting the user throughout the whole design process: code generation and rapid prototyping, random execution, interactive and guided simulation, model checking and equivalence checking, test case generation, and performance evaluation.

An LTS can be represented within CADP in two complementary ways: either *explicitly*, by its list of states and transitions encoded as a file in the BCG (*Binary Coded Graphs*) format equipped with specialized compression algorithms, or *implicitly*, by its successor function given as a C program complying to the interface defined by the OPEN/CÆSAR [7] environment for graph manipulation. The explicit representation is suitable for *global* verification algorithms, which explore transitions forward and backward, whereas the implicit representation is suitable for *local* (or *on-the-fly*) verification algorithms, which explore transitions forward, thus enabling an incremental construction of the LTS during verification. To deal with large systems, CADP provides several advanced analysis techniques: on-the-fly verification, partial order reductions, compositional verification, and massively parallel verification using clusters of machines.

CADP contains currently over 40 tools and libraries for LTS manipulation, which can be invoked either in interactive mode via the EUCALYPTUS graphical interface, or in batch mode via the SVL [8] scripting language dedicated to

Table 1. An extract of the process algebra formalizing BPEL, in discrete-time

BPEL	ATP
empty	empty $\xrightarrow{\surd} 0$
time	time $\xrightarrow{\chi}$ time
throw	$\forall e \in E_X$, throw$[e] \xrightarrow{e} 0$ with E_X set of exceptions that can be thrown.
receive / reply	$*o[m] \xrightarrow{*m}$ empty with $* \in \{?, !\}$ $*o[m] \xrightarrow{\chi} *o[m]$
sequence (;)	$\forall a \neq \surd, \ \dfrac{P \xrightarrow{a} P'}{P \ ; Q \xrightarrow{a} P' \ ; Q}$ $\forall a, \ \dfrac{P \xrightarrow{\surd} P' \ \wedge \ Q \xrightarrow{a} Q'}{P \ ; Q \xrightarrow{a} Q'}$
switch	switch$[\{P_i\}_{i \in I}]$ – $\forall i \in I$, switch$[\{P_i \mid i \in I\}] \xrightarrow{\tau} P_i$
while	while$[P] \xrightarrow{\tau} P$; while$[P]$ while$[P] \xrightarrow{\tau}$ empty
scope	Let $M_I = \{m_i \mid i \in I\}$ a set of messages and let $E_J = \{e_j \mid j \in J\}$ a set of exceptions. scope(P, E) with $E = [\{(m_i, P_i) \mid i \in I\}, (d, Q), \{(e_j, R_j) \mid j \in J\}]$ $\dfrac{P \xrightarrow{\surd}}{\text{scope}(P,E) \xrightarrow{\surd} 0}$ $\forall a \notin \{\chi, \surd\} \cup E_X \cup M_I \quad \dfrac{P \xrightarrow{a} P'}{\text{scope}(P,E) \xrightarrow{a} \text{scope}(P',E)}$ $d > 1, \ \dfrac{P \xrightarrow{\chi} P' \text{ and } \forall a \in E_X \cup \{\tau, \surd\}, \neg(P \xrightarrow{a})}{\text{scope}(P,E^d) \xrightarrow{\chi} \text{scope}(P,E^{d-1})}$ $\dfrac{P \xrightarrow{\chi} P' \text{ and } \forall a \in E_X \cup \{\tau, \surd\}, \neg(P \xrightarrow{a})}{\text{scope}(P,E^1) \xrightarrow{\chi} Q}$ $\forall i \in I, \ \dfrac{\forall a \in E_X \cup \{\tau, \surd\}, \neg(P \xrightarrow{a})}{\text{scope}(P,E) \xrightarrow{?m_i} P_i}$ $\forall j \in E_J, \ \dfrac{P \xrightarrow{e_j}}{\text{scope}(P,E) \xrightarrow{\tau} R_j}$ $\forall e \notin E_J, \ \dfrac{P \xrightarrow{e}}{\text{scope}(P,E) \xrightarrow{e} 0}$
pick	pick$[E]$ = scope(time, E) with $E = [\{(m_i, P_i) \mid i \in I\}, (d, Q), \{(e_j, R_j) \mid j \in J\}$

the description of complex verification scenarios. The toolbox was used for the validation of more than 100 industrial case-studies[1].

Since we focus on model checking discrete-time properties on DTLTSs, we could apply in principle existing tools operating on LTSs, such as the EVALUATOR 3.5 [20] on-the-fly model checker of CADP, which takes as input temporal formulas expressed in regular alternation-free μ-calculus. However, the evaluation of discrete-time properties requires the counting of *time* actions in the DTLTS; this can be encoded in μ-calculus using fixed point operators (one operator for each counter value), but may lead to prohibitively large temporal formulas, as noticed in the framework of temporal CCS [23]. Discrete-time properties can be

[1] See the online catalog at `http://www.inrialpes.fr/vasy/cadp/case-studies`

naturally formulated using data-handling extensions of the modal μ-calculus, such as the MCL language [21] underlying the EVALUATOR 4.0 tool recently integrated into CADP. We will illustrate the usage of MCL in Section 3.3.

3 Case Study: A Web Service for GPS Navigation

We illustrate in this section the application of our approach to the modeling and analysis of a Web service dedicated to GPS navigation. Given the relative complexity of this Web service, we do not detail here its textual BPEL and WSDL descriptions, but present its workflow graphically using the BPMN [10] notation.

3.1 System Description

The purpose of the GPS Web service is to compute itineraries from a position to a destination fixed by a user (client of the service). In addition to the requested itinerary, the user can also obtain: pictures of the travel (taken from the air), the global map of the itinerary, and various kinds of information (about traffic, radar stations, point of interest (POI), etc). At last, the user can configure the subscription to the various kinds of information, as well as some parameters of the travel (e.g., to take motorway or not, to deviate toward a POI, etc.). The relationships between these functionalities are represented in Figure 3.

The behavior of the GPS Web service consists of two main phases, described in the sequel: the initialization phase (login, setting of the initial position and destination) and the main loop phase (management of the itinerary, modification of the parameters, etc.).

Initialization Phase. The initialization phase comprises three activities: login, position and destination.

Login activity. The access to the Web service is restricted to authenticated users only. To identify itself, the user must send a couple login/password, to which the service responds by a message "Ok" or "NOk" depending whether the couple is valid or not.

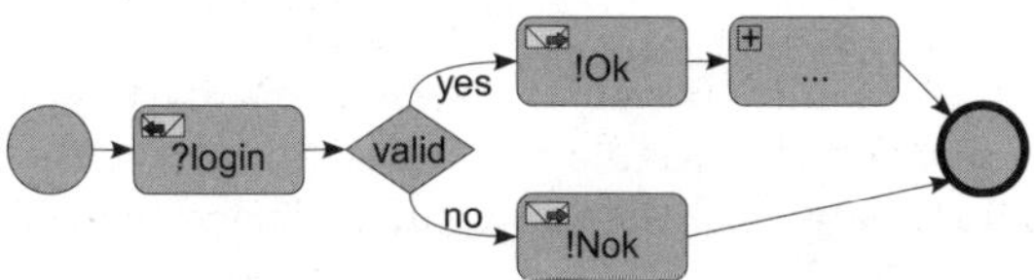

Position activity. After authentication, in order to use the main functionalities of the Web service, the user must indicate where the start location of the travel is. This is done by sending a message with information about the street, city, and country where the navigation session must be started; the message must be resent until the start location is accepted by the service (message "Ok").

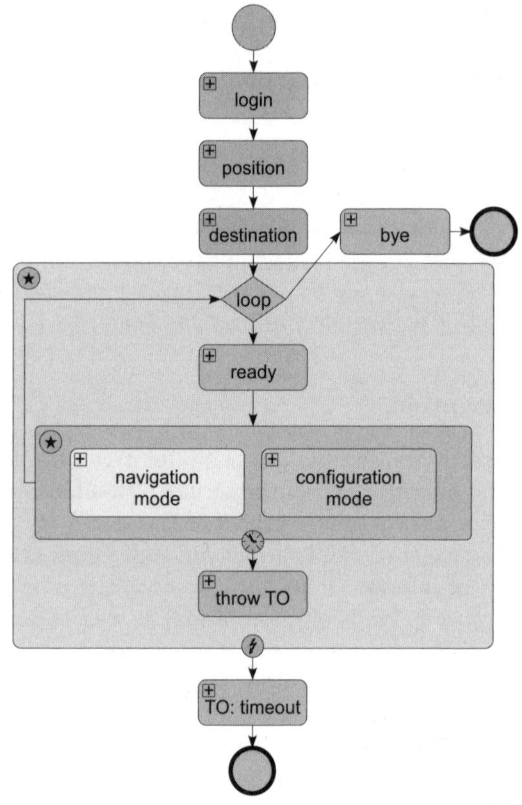

Fig. 3. Functionality workflow of the GPS Web service

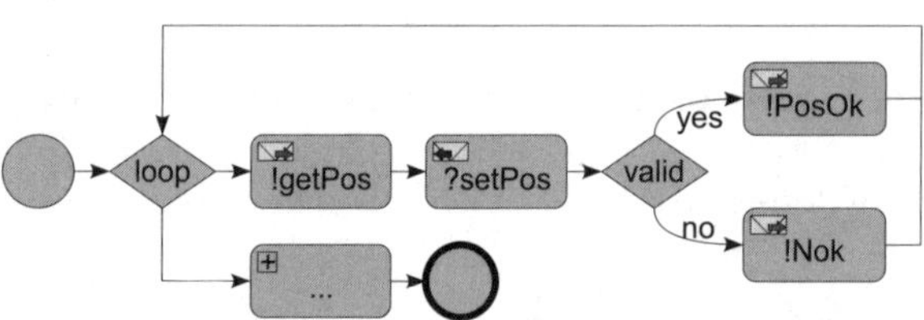

Destination activity. Finally, before the service may attempt to calculate an itinerary, the user must enter a destination. This is similar to the position activity above: the user must retry until the end location is accepted by the service.

Main Loop Phase. After the initialization phase, the service can compute an itinerary, send information about traffic, POI, etc. To maintain the connection with the user, the service requires that the time elapsed between certain consecutive user actions does not exceed a given timeout value (a kind of "ping alive"). In Section 3.2 we will consider for analysis configurations of the system with a timeout value ranging from 1 to 60 seconds.

From the Web service point of view (see Figure 3), this timeout is managed by a "scope" process: when the timeout is reached, this process generates an exception that will be caught by another process. The main activity of this "scope" process is a "pick" process. This kind of choice enables the user to select a desired action; if we used the "switch" BPEL construct instead, the choice would be made by the Web service and not by the user. Furthermore, the "scope" is encapsulated into a "while", enabling the user to do more than one operation during the session (notice that the first action carried out by the service when entering the "while" is the emission of a *ready* message to the user). Finally, the "while" is the main activity of a second "scope", that catches the first exception thrown when the timeout is reached.

The activities executed by the main loop are partitioned in two modes, described in the sequel: the navigation mode (obtaining the itinerary, modifying the current position or destination, getting a picture or a roadmap), and the configuration mode (subscribing to a POI, getting information on radars or traffic, setting of parameters).

Navigation mode. In navigation mode, the user can change the current position and the destination (using the same procedure as for the initialization phase). Next, the user can ask for the itinerary, a picture, the roadmap, or enter in configuration mode. There are two types of answer for itinerary requests: either a complete itinerary leading from the current position to the destination, with various information (about street, radar, POI, etc.) depending on the user subscriptions, or simply a *destination* message indicating that the current position is (near) to the destination. The requests for picture and roadmap allow the user to obtain an air-picture of the area (in PNG format) or a veritable roadmap (in SVG format).

Configuration mode. In configuration mode, the user can subscribe or cancel his subscription to information about POI, radar or traffic. This information is added to the itinerary if necessary. Additionally, the user can set various parameters, such as the kind of the itinerary (on motorway or not), etc.

3.2 Discrete-Time LTS Synthesis

Starting from the BPEL description of the GPS Web service, we apply the WSMOD tool in order to obtain a DTLTS on which the verification tools of CADP will operate. We show below the DTLTS model obtained for a timeout of 1 second, then we study its variation in size as the timeout value increases, and finally we discuss the behavior of the Web service w.r.t. the ambiguity detection feature implemented in WSMOD.

Discrete-timed labeled transition system. DTLTS models represent the observable behavior of Web services. The actions labeling the DTLTS transitions denote the messages exchanged (emissions and receptions are prefixed by '!' and '?', respectively), the elapse of a discrete-time unit χ (or *time*), the internal action

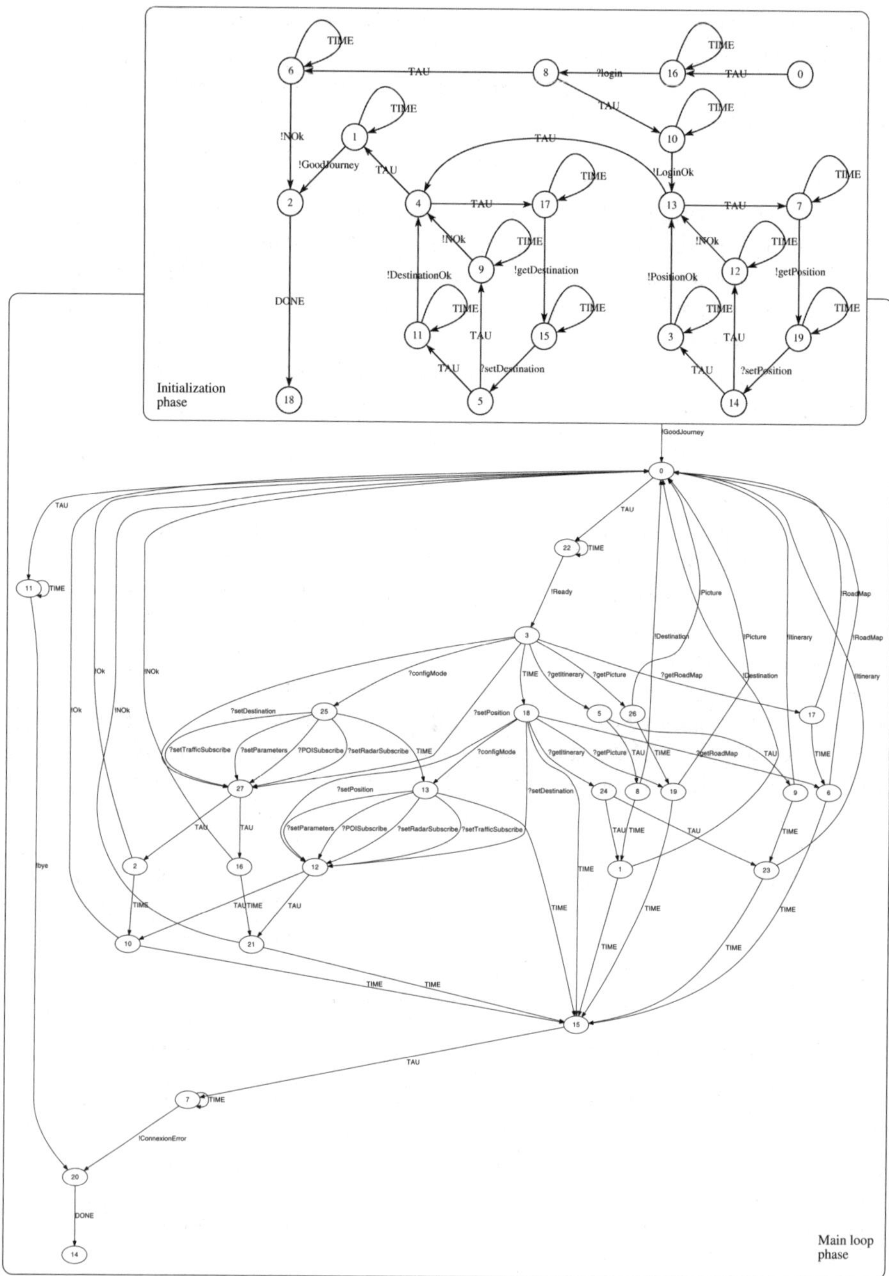

Fig. 4. DTLTS model of the GPS Web service, with zoom on the initialization phase. The action *!GoodJourney* makes the link between the initialization phase and the main loop phase.

τ (or *tau*), and the termination action $\surd$ (or *done*). The global behavior of the Web service is obtained by an exhaustive simulation of the BPEL description driven by the ATP rules given in Table 1. The DTLTS obtained in this manner for the GPS Web service with a timeout value of 1 second is shown in Figure 4.

Variation of DTLTS size with the timeout value. The size of the DTLTS (number of states and transitions) depends on several aspects of the BPEL description: the number of BPEL processes, their complexity and nesting, the amount of communications, and the values of the timeouts. For the sake of readability, we have shown in Figure 4 the DTLTSs for a timeout of 1 second (corresponding to one χ in discrete-time), but we carried out verification also for larger values of the timeout.

The figure on the right shows the variation of the DTLTS size for timeout values ranging from 1 to 60. We observe a linear increase of both the number of states and transitions; this is a consequence of the fact that the BPEL description contains a single timeout (according to the ATP rules). In the presence of multiple, overlapped timeouts, the size of the DTLTS may increase much more quickly.

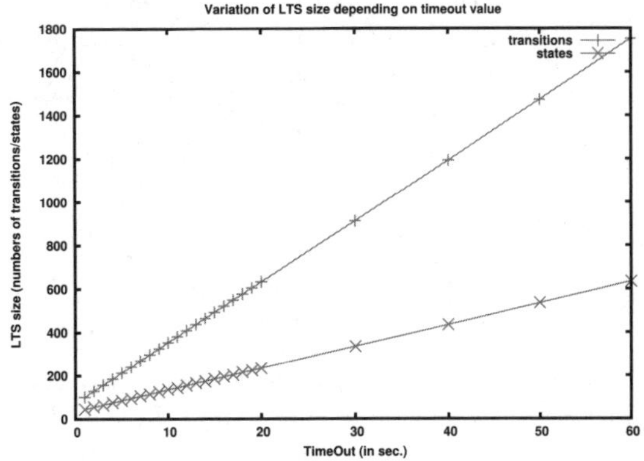

Non ambiguous Web service. In this study, we focus on the verification of the Web service behavior. However, the WSMOD tool can also synthesize automatically a DTLTS modeling the behavior of an adapted client interacting with the Web service, provided that the model of the service respects certain properties (concerning non ambiguity in message exchanges, time elapsed, etc.) detailed in [11]. Here, the GPS Web service is identified as *non ambiguous* by WSMOD, meaning that the tool can synthesize an adapted client that can know, on each message exchange, the exact choice made on the service side, and therefore the client and the service can evolve without any deadlocks or mismatches.

3.3 Verification of Discrete-Time Properties

We analyze below the behavior of the GPS Web service (considering a timeout of 50 seconds) by means of discrete-time model checking using the EVALUATOR 4.0 [21] tool of CADP. Table 2 illustrates the formulation in MCL of several safety and liveness properties, of both untimed and timed nature. The colored parts of the formulas indicate discrete-time properties, which involve the counting of *time* actions. All properties were successfully verified on the corresponding DTLTS of the system, which has 535 states and 1473 transitions.

Table 2. Safety and liveness properties of the GPS Web service (timeout of 50 sec.)

Prop.	MCL formula
S_1	$[\ (\neg !LoginOk)^*.?setPosition \vee ?setDestination\ ]$ false
S_2	$[\ (\mathsf{true}^*.$ $((!getPosition.(\neg ?setPosition)^*) \mid (!getDestination.(\neg ?setDestination)^*))\ .$ $!GoodJourney)\ ]$ false
S_3	$[\ \mathsf{true}^*.?getItinerary.(\neg(!Itinerary \vee !Destination))^*.$ $(?getPicture \vee ?getRoadMap \vee ?configMode \vee$ $?setPosition \vee ?setDestination \vee ?getItinerary)\ ]$ false
S_4	$[\ \mathsf{true}^*.?getItinerary.(\neg(!Itinerary \vee !Destination))^*.$ $(time.(\neg(!Itinerary \vee !Destination))^*)\{51\}\ .$ $(!Itinerary \vee !Destination)\ ]$ false
L_1	$[\ \mathsf{true}^*.!LoginOk\ ]$ AF $\langle !getPosition \vee !getDestination \vee !GoodJourney \rangle$ true
L_2	$[\ \mathsf{true}^*.!GoodJourney.(\tau \vee time)^*\ ]\ \langle (\tau \vee time)^*.\ !Ready \vee !bye \rangle$ true)
L_3	$[\ \mathsf{true}^*.!Ready.\ time\{\ ...\ 50\ \}\ ]$ $\langle \mathsf{true}^*.!Picture \vee !RoadMap \vee !Itinerary \vee !Destination \rangle$ true
L_4	$[\ \mathsf{true}^*.!Ready.$ $((\neg(!Itinerary \vee !Destination \vee !Picture \vee !RoadMap))^*.time)\{51\}\ ]$ AF $\langle !ConnectionError \rangle$ true

Safety properties: they specify informally that "something bad never happens" during the execution of the system. In the MCL language, these properties can be expressed in a concise manner by identifying the undesirable execution sequences, characterizing them using extended regular expressions, and forbidding their existence in the DTLTS model using necessity modalities.

Properties S_1 and S_2 concern the ordering of actions during the initialization phase: S_1 specifies that the user cannot set the position or the destination before logging in successfully, and S_2 states that after requesting the position or the destination, the Web service cannot begin the main loop before receiving an appropriate answer from the user. Properties S_3 and S_4 deal with the main loop phase: S_3 forbids the user to make another request before the current one (here, an itinerary demand) has been handled by the service, and S_4 states that a demand cannot be fulfilled anymore by the service after the timeout has expired. The $R\{...\,n\}$ and $R\{n\}$ extended regular operators denote the repetition of a regular expression R at most n times and exactly n times, respectively.

Liveness properties: they specify informally that "something good eventually happens" during the execution of the system. In MCL, these properties contain diamond modalities and minimal fixed point operators for encoding the existence of certain desirable execution sequences (*potentiality*) or trees (*inevitability*) in the DTLTS.

Properties L_1 and L_2 concern the initialization phase: L_1 specifies that after the user has logged in, the Web service will eventually ask for the position, the destination, or end the initialization, and L_2 states that after the initialization was finished the service will end up in the main loop or decide to terminate the session. Properties L_3 and L_4 deal with the main loop phase: L_3 indicates that as long as the timeout has not expired, the service can still prompt for a user request, and L_4 states that an expiration of the timeout eventually interrupts the connection. The $\mathsf{AF}\ p$ operator of CTL [3] expressing the inevitable reachability of a state p is defined in μ-calculus as $\mu X.p \vee (\langle \mathsf{true} \rangle\, \mathsf{true} \wedge [\mathsf{true}]X)$.

4 Conclusion and Future Work

The design of complex business processes according to the SOA approach requires to carefully take into account the presence of concurrency, communication, and timing constraints induced by the interaction of Web services. To facilitate the design process, we propose here a tool-equipped methodology for modeling and analyzing Web services described in BPEL. We focus on the behavioral and discrete-time aspects of Web services, and rely upon the model-based verification technologies stemming from the concurrency domain. The state/transition models of BPEL Web services are produced automatically by the WSMOD tool, which implements an exhaustive simulation algorithm based on a formalization of BPEL semantics by means of process algebraic rules. The tool is able to handle both discrete and continuous time constraints; for the moment we handle only discrete-time models, which can be analyzed using the EVALUATOR 4.0 model checker [21] of the CADP toolbox [9]. Discrete-time safety and liveness properties can be concisely expressed using the data-handling facilities of the MCL language accepted as input by EVALUATOR 4.0, and particularly the extended regular expressions over transition sequences, which allow to count tick actions occurring in the model. We illustrated the verification of discrete-time properties on the example of a GPS Web service; however, most of them can be easily adapted for other business processes described in BPEL. Our methodology enables the Web service designers to carry out formal analysis on complex Web services before publishing them, and thus to improve the quality of the design process.

We plan to continue our work along several directions. Firstly, we can improve the connection between WSMOD and CADP by producing implicit DTLTSs according to the interface defined by OPEN/CÆSAR [7]. This would enable on-the-fly verification, which allows to detect errors in large systems without constructing the complete DTLTS model beforehand but exploring it in a demand-driven way. Secondly, using discrete-time models allows to directly reuse the tools available for data-based temporal logics, such as EVALUATOR 4.0; however, this may lead to state explosion in the presence of numerous timeouts. An alternative solution would be to use continuous time models; this can be achieved by connecting the time automata produced by WSMOD with the UPPAAL [17] tool dedicated to the verification of continuous time models. Finally, we will extend the methodology to handle compositions of multiple Web services, following our previous work on automated client synthesis [22], but focusing on the verification of composition. For this purpose, the compositional verification techniques available in tools such as EXP.OPEN [16] will be certainly useful.

References

1. Alur, R., Dill, D.L.: A theory of timed automata. Theoretical Computer Science 126(2), 183–235 (1994)
2. Chirichiello, A., Salaün, G.: Encoding abstract descriptions into executable web services: Towards a formal development. In: Proc. of WI 2005, pp. 457–463. IEEE Computer Society, Los Alamitos (2005)
3. Clarke, E., Grumberg, O., Peled, D.: Model Checking. MIT Press, Cambridge (2000)
4. Ferrara, A.: Web services: a process algebra approach. In: ICSOC, pp. 242–251 (2004)
5. Fu, X., Bultan, T., Su, J.: Analysis of interacting BPEL web services. In: Proc. of the 13th International World Wide Web Conference (WWW 2004), USA, ACM Press, New York (2004)
6. Fu, X., Bultan, T., Su, J.: WSAT: A tool for formal analysis of web services. In: Alur, R., Peled, D.A. (eds.) CAV 2004. LNCS, vol. 3114, Springer, Heidelberg (2004)
7. Garavel, H.: OPEN/CÆSAR: An open software architecture for verification, simulation, and testing. In: Steffen, B. (ed.) ETAPS 1998 and TACAS 1998. LNCS, vol. 1384, pp. 68–84. Springer, Heidelberg (1998)
8. Garavel, H., Lang, F.: SVL: a scripting language for compositional verification. In: Proc. of FORTE 2001, IFIP, pp. 377–392. Kluwer Academic Publishers, Dordrecht (2001); Full version available as INRIA Research Report RR-4223
9. Garavel, H., Lang, F., Mateescu, R., Serwe, W.: CADP 2006: A toolbox for the construction and analysis of distributed processes. In: Damm, W., Hermanns, H. (eds.) CAV 2007. LNCS, vol. 4590, pp. 158–163. Springer, Heidelberg (2007)
10. Object Management Group. Business process modeling notation (BPMN) specification (May 2006)
11. Haddad, S., Melliti, T., Moreaux, P., Rampacek, S.: Modelling web services interoperability. In: Proc. of the 6th Int. Conf. on Enterprise Information Systems (ICEIS 2004), Porto, Portugal (April14–17, 2004)
12. Haddad, S., Moreaux, P., Rampacek, S.: A formal semantics and a client synthesis for a BPEL service. In: ICEIS 2006, Revised Selected Paper. Lecture Notes in Business Information Processing, vol. 3, Springer, Heidelberg (2008)
13. ISO/IEC. LOTOS — a formal description technique based on the temporal ordering of observational behaviour. International Standard 8807, International Organization for Standardization — Information Processing Systems — Open Systems Interconnection, Genève (September 1989)
14. Jordan, D., Evdemon, J.: Web Services Business Process Execution Language Version 2.0 - Oasis Standard (April 11, 2007)
15. Josuttis, N.: Soa in Practice – The Art of Distributed System Design, O'Reilly Media, City (2007)
16. Lang, F.: EXP.OPEN 2.0: A flexible tool integrating partial order, compositional, and on-the-fly verification methods. In: Romijn, J.M.T., Smith, G.P., van de Pol, J. (eds.) IFM 2005. LNCS, vol. 3771, Springer, Heidelberg (2005)
17. Larsen, K.G., Pettersson, P., Yi, W.: UPPAAL in a Nutshell. Int. Journal on Software Tools for Technology Transfer 1(1–2), 134–152 (1997)
18. Magee, J., Kramer, J.: Concurrency: State Models and Java Programs. Wiley, Chichester (1999)

19. Martin, A.J.: Compiling communicating processes into delay-insensitive VLSI circuits. Distributed Computing 1(4), 226–234 (1986)
20. Mateescu, R., Sighireanu, M.: Efficient on-the-fly model-checking for regular alternation-free mu-calculus. Science of Computer Programming 46(3), 255–281 (2003)
21. Mateescu, R., Thivolle, D.: A model checking language for concurrent value-passing systems. In: Proc. of FM 2008. LNCS, vol. 5014, Springer, Heidelberg (2008)
22. Melliti, T., Boutrous-Saab, C., Rampacek, S.: Verifying correctness of web services choreography. In: Proc. of ECOWS 2006, Zurich, Switzerland, IEEE Computer Society Press, Los Alamitos (2006)
23. Morley, M.J.: Safety-level communication in railway interlockings. Science of Computer Programming 29(1-2), 147–170 (1997)
24. Nicollin, X., Sifakis, J.: The algebra of timed processes ATP: Theory and application (1994)
25. Salaün, G., Ferrara, A., Chirichiello, A.: Negotiation Among Web Services Using LOTOS/CADP. In: (LJ) Zhang, L.-J., Jeckle, M. (eds.) ECOWS 2004. LNCS, vol. 3250, pp. 198–212. Springer, Heidelberg (2004)
26. Salaün, G., Kramer, J., Lang, F., Magee, J.: Translating FSP into LOTOS and networks of automata. In: Davies, J., Gibbons, J. (eds.) IFM 2007. LNCS, vol. 4591, pp. 558–578. Springer, Heidelberg (2007)
27. Salaün, G., Serwe, W.: Translating hardware process algebras into standard process algebras — illustration with CHP and LOTOS. In: Proc. of IFM 2005. LNCS, vol. 3371, pp. 287–306. Springer, Heidelberg (2005)
28. Turner, K.J.: Representing and analysing composed web services using CRESS. J. Netw. Comput. Appl. 30(2), 541–562 (2007)

Author Index

Printing: Mercedes-Druck, Berlin
Binding: Stein+Lehmann, Berlin